Cram101 Textbook Outlines to accompany:

Marine Biology: Function, Biodiversity, Ecology

Levinton, 2nd Edition

An Academic Internet Publishers (AIPI) publication (c) 2007.

With Cram101.com online, you also have access to extensive reference material.

You will nail those essays and papers. Here is an example from a Cram101 Biology text:

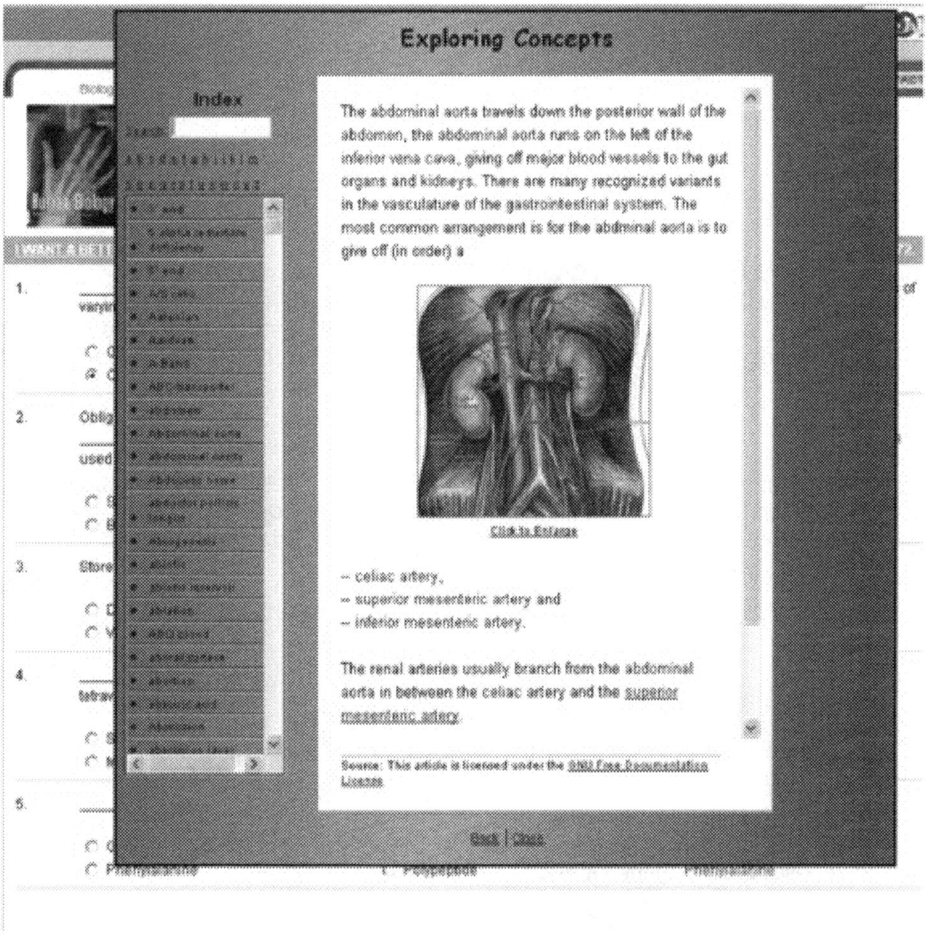

Visit **www.Cram101.com**, click Sign Up at the top of the screen, and enter DK73DW in the promo code box on the registration screen. Access to www.Cram101.com is normally $9.95, but because you have purchased this book, your access fee is only $4.95. Sign up and stop highlighting textbooks forever.

Learning System

Cram101 Textbook Outlines is a learning system. The notes in this book are the highlights of your textbook, you will never have to highlight a book again.

How to use this book. Take this book to class, it is your notebook for the lecture. The notes and highlights on the left hand side of the pages follow the outline and order of the textbook. All you have to do is follow along while your intructor presents the lecture. Circle the items emphasized in class and add other important information on the right side. With Cram101 Textbook Outlines you'll spend less time writing and more time listening. Learning becomes more efficient.

Cram101.com Online

Increase your studying efficiency by using Cram101.com's practice tests and online reference material. It is the perfect complement to Cram101 Textbook Outlines. Use self-teaching matching tests or simulate in-class testing with comprehensive multiple choice tests, or simply use Cram's true and false tests for quick review. Cram101.com even allows you to enter your in-class notes for an integrated studying format combining the textbook notes with your class notes.

Visit **www.Cram101.com**, click Sign Up at the top of the screen, and enter **DK73DW1906** in the promo code box on the registration screen. Access to www.Cram101.com is normally $9.95, but because you have purchased this book, your access fee is only $4.95. Sign up and stop highlighting textbooks forever.

 ISBN: 1-4288-0366-1

Marine Biology: Function, Biodiversity, Ecology
Levinton, 2nd

CONTENTS

Crab	The term crab is often applied to several different groups of short (nose to tail) decapod crustaceans with thick exoskeletons, but only members of the Brachyura are true.
Microscope	A microscope is an instrument for viewing objects that are too small to be seen by the naked or unaided
Protozoa	Protozoa are single-celled eukaryotes (organisms with nuclei) that show some characteristics usually associated with animals, most notably mobility and heterotrophy. They are often grouped in the kingdom Protista together with the plant-like algae.
Biology	Biology is the branch of science dealing with the study of life. It is concerned with the characteristics, classification, and behaviors of organisms, how species come into existence, and the interactions they have with each other and with the environment.
Ecology	Ecology is the scientific study of the distribution and abundance of living organisms and how these properties are affected by interactions between the organisms and their environment.
Coral reef	A coral reef is a type of biotic reef that develops in tropical waters by the growth of coralline algae, hermatypic corals, and other marine organisms.
Element	A chemical element, often called simply element, is a chemical substance that cannot be divided or changed into other chemical substances by any ordinary chemical technique. An element is a class of substances that contain the same number of protons in all its atoms.
Reproduction	Biological reproduction is the biological process by which new individual organisms are produced. Reproduction is a fundamental feature of all known life; each individual organism exists as the result of reproduction by an antecedent.
Respiration	Respiration is the process by which an organism obtains energy by reacting oxygen with glucose to give water, carbon dioxide and ATP (energy). Respiration takes place on a cellular level in the mitochondria of the cells and provide the cells with energy.
Metabolism	Metabolism is the biochemical modification of chemical compounds in living organisms and cells. This includes the biosynthesis of complex organic molecules (anabolism) and their breakdown (catabolism).
Digestion	Digestion refers to the mechanical and chemical breakdown of food into molecules small enough for the body to absorb; the second main stage of food processing, following ingestion.
Distribution	Distribution in pharmacology is a branch of pharmacokinetics describing reversible transfer of drug from one location to another within the body.
Biodiversity	Biodiversity is the totality of genes, species, and ecosystems of a region.
Deep sea	The deep sea refers to those areas of oceans to which little or no light penetrates.
Seaweed	Phycologists consider seaweed to refer any of a large number of marine benthic algae that are multicellular, macrothallic (large-bodied), and thus differentiated from most algae that tend to be microscopic in size.
Predator	A predator is an animal or other organism that hunts and kills other organisms for food in an act called predation.
Habitat	Habitat refers to a place where an organism lives; an environmental situation in which an organism lives.
Anatomy	Anatomy is the branch of biology that deals with the structure and organization of living things. It can be divided into animal anatomy (zootomy) and plant anatomy (phytonomy).
Territory	In ethology, sociobiology and behavioral ecology, the term territory refers to any geographical area that an animal of a particular species consistently defends against conspecifics (and, occasionally, animals of other species).
Mollusk	The mollusk is the large and diverse phylum Mollusca, which includes a variety of familiar creatures

	well-known for their decorative shells or as seafood. These range from tiny snails, clams, and abalone to the octopus, cuttlefish and squid (which are considered the most intelligent invertebrates).
Species	Group of similarly constructed organisms capable of interbreeding and producing fertile offspring is a species.
Natural selection	Natural selection is the process by which biological individuals that are endowed with favorable or deleterious traits end up reproducing more or less than other individuals that do not possess such traits.
Charles Darwin	Charles Darwin was a British naturalist who achieved lasting fame as originator of the theory of evolution through natural selection.
Evolution	In biology, evolution is the process by which novel traits arise in populations and are passed on from generation to generation. Its action over large stretches of time explains the origin of new species and ultimately the vast diversity of the biological world.
Barnacle	A barnacle is a type of arthropod belonging to infraclass Cirripedia in the subphylum Crustacea and is hence distantly related to crabs and lobsters. They live attached to surfaces and are typically enclosed by heavy calcareous plates are referred to as barnacles.
Colonization	Colonization is the act where life forms move into a distant area where their kind is sparse or not yet existing at all and set up new settlements in the area. Colonization applies to all life forms in a sense though it is most often used in reference to insects and humans.
Coral	Group of cnidarians having a calcium carbonate skeleton that participate in the formation of reefs is coral.
Atoll	An atoll is a type of low, coral island found in tropical oceans and consisting of a coral-algal reef surrounding a central depression. The depression may be part of the emergent island, but more typically is a part of the sea (that is, a lagoon), or very rarely is an enclosed body of fresh, brackish, or highly saline water.
Subsidence	Subsidence is a term used in geology, engineering and surveying to denote the motion of a surface (usually, the earth's surface) downwards relative to a datum such as sea-level.
Alcohol	Alcohol is a general term, applied to any organic compound in which a hydroxyl group (-OH) is bound to a carbon atom, which in turn is bound to other hydrogen and/or carbon atoms. The general formula for a simple acyclic alcohol is $C_nH_{2n+1}OH$.
Embryology	Embryology is the branch of developmental biology that studies embryos and their development
Starfish	Starfish are marine invertebrates belonging to phylum Echinodermata, class Asteroidea. They exhibit a superficially radial symmetry, typically with five or more "arms" which radiate from an indistinct disk. They do not have movable skeletons, but instead possess a hydraulic water vascular system.
Plankton	Plankton are drifting organisms that inhabit the water column of oceans, seas, and bodies of fresh water. Plankton abundance and distribution are strongly dependent on factors such as ambient nutrients concentrations, the physical state of the water column, and the abundance of other plankton.
Biota	Biota is the plant and animal life of a region or area or geological period. The biota of the Earth is known as the biosphere.
Chlorophyll	Chlorophyll is a green photosynthetic pigment found in plants, algae, and cyanobacteria. In plant photosynthesis incoming light is absorbed by chlorophyll and other accessory pigments in the antenna complexes of photosystem I and photosystem II.
Climate	Weather condition of an area including especially prevailing temperature and average daily/yearly rainfall over a long period of time is called climate.
Scientific	A way of gaining information about the world around you that involves observation, hypothesis

method	formation, testing of hypotheses, theory formation, and law formation is called scientific method.
Prediction	Step of the scientific method which follows the formulation of a hypothesis and assists in creating the experimental design is referred to as prediction.
Inspiration	Inspiration begins with the onset of contraction of the diaphragm, which results in expansion of the intrapleural space and an increase in negative pressure according to Boyle's Law.
Salt	Salt is a term used for ionic compounds composed of positively charged cations and negatively charged anions, so that the product is neutral and without a net charge.
Migration	Migration occurs when living things move from one biome to another. In most cases organisms migrate to avoid local shortages of food, usually caused by winter. Animals may also migrate to a certain location to breed, as is the case with some fish.
Oxygen	Oxygen is a chemical element in the periodic table. It has the symbol O and atomic number 8. Oxygen is the second most common element on Earth, composing around 46% of the mass of Earth's crust and 28% of the mass of Earth as a whole, and is the third most common element in the universe.
Population	Group of organisms of the same species occupying a certain area and sharing a common gene pool is referred to as population.
Null hypothesis	Null hypothesis refers to a statistical hypothesis to be tested and either accepted or rejected in favor of an alternative.
Predation	Interaction in which one organism uses another, called the prey, as a food source is referred to as predation.
Population density	The number of organisms of a species per unit area is called population density.
Threshold	Electrical potential level at which an action potential or nerve impulse is produced is called threshold.
Benthos	Benthos are the organisms and habitats of the sea floor; in freshwater biology they are the organisms and habitats of the bottoms of lakes, rivers, and creeks.
Neuston	Neuston refers to small- to medium-size organisms that live on the surface of the sea.
Nekton	Animals, such as fish, that can swim against currents of water are referred to as nekton.
Microorganism	A microorganism is an organism that is so small that it is microscopic (invisible to the naked eye). They are often illustrated using single-celled, or unicellular organisms; however, some unicellular protists are visible to the naked eye, and some multicellular species are microscopic.
Water column	The vertical column of seawater that extends from the surface to the bottom is called water column.
Intertidal zone	The part of a beach that lies between average high tide and average low tide is called the intertidal zone.
Tide	The periodic, rhythmic rise and fall of the sea surface caused by changes in gravitational forces external to the Earth is referred to as the tide.
Subtidal zone	The subtidal zone is the area below the range of tidal exposure; always covered by water.
Continental shelf	The shallow part of the seafloor immediately adjacent to a continent is called the continental shelf.
Abyssopelagic	The pelagic zone is the part of the open sea or ocean comprising the water column, i.e. all of the sea other than that near the coast or the sea floor. Abyssopelagic (from 4000 metres down to above the ocean floor) - no light whatsoever penetrates to this depth, and most creatures are blind and colorless.

Bathypelagic	Bathypelagic is from 1000 m down to around 4000 m, by this depth the ocean is almost entirely dark. There are no living plants, and most animals survive by consuming the snow of detritus falling from the zones above, or by preying upon others. Giant squid live at this depth, and here they are hunted by deep-diving sperm whales.
Mesopelagic	The pelagic zone is the part of the open sea or ocean comprising the water column, i.e. all of the sea other than that near the coast or the sea floor. Mesopelagic (from 200 metres down to around 1000 metres) - the twilight zone. Although some light penetrates this deep, it is insufficient for photosynthesis.
Epipelagic	Epipelagic (from the surface down to around 200 metres) - the illuminated surface zone where there is enough light for photosynthesis, and thus plants and animals are largely concentrated in this zone. Here one will typically encounter fish such as tuna and many sharks.
Temperate	In virology, temperate refers to the life cycle a phage is able to perform. The phage usually integrates its genome into its host chromosome, becoming a prophage.

Climate	Weather condition of an area including especially prevailing temperature and average daily/yearly rainfall over a long period of time is called climate.
Sunlight	Sunlight in the broad sense is the total spectrum of electromagnetic radiation given off by the Sun.
Salt	Salt is a term used for ionic compounds composed of positively charged cations and negatively charged anions, so that the product is neutral and without a net charge.
Continental shelf	The shallow part of the seafloor immediately adjacent to a continent is called the continental shelf.
Continental slope	Continental slope the part of the continental margin that lies between the continental shelf and the bottom of the ocean. Sunlight does not penetrate this area, and mostly it is home to scavengers. It is characterized by a relatively steep slope of 3° to 6°.
Submarine canyon	A Submarine canyon is a steep-sided valley on the seafloor of the continental slope. Submarine canyons are generally found as extensions to large rivers, and have been found to extend 1km below sea level, and extend for hundreds of kilometers.
Sediment	Sediment is any particulate matter that can be transported by fluid flow and which eventually is deposited as a layer of solid particles on the bed or bottom of a body of water or other liquid.
Continental rise	The gently sloping area at the base of the continental slope is called the continental rise.
Abyssal plain	Abyssal plain is a flat or very gently sloping area of the deep ocean basin floor. They are among the Earth's flattest and smoothest regions and the least explored. They cover approximately 40% of the ocean floor and reach depths between 2,200 and 5,500 m). They generally lie between the foot of a continental rise and a mid-oceanic ridge.
Continental island	Continental island refers to an island that is near to and geologically part of a continent, for example, the British Isles or Trinidad.
Transverse	A transverse (also known as axial or horizontal) plane is an X-Y plane, parallel to the ground, which (in humans) separates the superior from the inferior, or put another way, the head from the feet.
Crust	A crust is the outer layer of a planet, part of its lithosphere. Planetary crust is generally composed of a less dense material than that of its deeper layers. The crust of the Earth is composed mainly of basalt and granite.
Snake	A snake also known as ophidians, are cold blooded legless reptiles closely related to lizards, which share the order Squamata.
Manganese nodule	Manganese nodule is a rock concretion on the sea bottom formed of concentric layers of iron and manganese hydroxides around a core. They are one of the slowest of all geological phenomena – in the order of a centimeter over several million years.
Skeleton	In biology, the skeleton or skeletal system is the biological system providing physical support in living organisms.
Plankton	Plankton are drifting organisms that inhabit the water column of oceans, seas, and bodies of fresh water. Plankton abundance and distribution are strongly dependent on factors such as ambient nutrients concentrations, the physical state of the water column, and the abundance of other plankton.
Clay	A soil constituent comprising particles smaller than 2 micrometers in diameter is clay. Clay is a generic term for an aggregate of hydrous silicate particles less than 2 ìm (micrometres) in diameter.

Calcium carbonate	A mineral that is the major component of the shell, skeleton, and other parts of many organisms are called calcium carbonate.
Radiolarian ooze	Radiolarian ooze occurrs in many places in very deep water it is composed mainly of the siliceous skeletons of radiolarians, and calcareous matter that's dissolved by the lage percentage of carbon dioxide in the water at these depths.
Foraminifera	The Foraminifera, or forams for short, are a large group of amoeboid protists with reticulating pseudopods, fine strands that branch and merge to form a dynamic net. They typically produce a shell, or test, which can have either one or multiple chambers, some becoming quite elaborate in structure.
Silica	Silica is a mineral similar to glass that is the major component of the cell wall, shell, or skeleton of many marine organisms.
Sedimentary rock	Sedimentary rock is one of the three main rock groups and is formed in three main ways—by the deposition of the weathered remains of other rocks; by the deposition of the results of biogenic activity; and by precipitation from solution.
Polarity	In cell biology, polarity refers to cells not being point-symmetrical in their spatial organization. In horticulture, polarity refers to the condition in which cuttings grow shoots at the distil end and roots at the proximal end.
Magnetic anomalies	Magnetic anomalies refers to magnetic bands in the sea floor that run parallel to the mid-ocean ridge .
Mantle	The mantle is an organ found in mollusks. It is the dorsal body wall covering the main body, or visceral mass. The epidermis of this organ secretes calcium carbonate to create a shell.
Seafloor spreading	Seafloor spreading is a part of the theory of plate tectonics. Seafloor spreading is the process by which continental drift occurs. This mechanism, which is a more accurate version of Alfred Wegener's original drift of continents that "plow" through the sea.
Radioactive	A term used to describe the property of releasing energy or particles from an unstable atom is called radioactive.
Radioactive decay	Radioactive decay is the set of various processes by which unstable atomic nuclei (nuclides) emit subatomic particles (radiation).
Continental drift	Continental drift refers to movement of continents with respect to one another over the earth's surface.
Heat of vaporization	The energy that must be supplied to a compound to transform it from a liquid into a gas at its boiling temperature is called heat of vaporization.
Specific heat	The amount of energy required to raise the temperature of 1 gram of a substance by 1 °C is specific heat.
Hydrogen bond	A hydrogen bond is a type of attractive intermolecular force that exists between two partial electric charges of opposite polarity. Although stronger than most other intermolecular forces, the typical hydrogen bond is much weaker than both the ionic bond and the covalent bond.
Solvent	A solvent is a liquid that dissolves a solid, liquid, or gaseous solute, resulting in a solution. The most common solvent in everyday life is water.
Ion	Ion refers to an atom or molecule that has gained or lost one or more electrons, thus acquiring an electrical charge.
Hydrogen	Hydrogen is a chemical element in the periodic table that has the symbol H and atomic number 1. At standard temperature and pressure it is a colorless, odorless, nonmetallic, univalent,

	tasteless, highly flammable diatomic gas.
Oxygen	Oxygen is a chemical element in the periodic table. It has the symbol O and atomic number 8. Oxygen is the second most common element on Earth, composing around 46% of the mass of Earth's crust and 28% of the mass of Earth as a whole, and is the third most common element in the universe.
Atom	An atom is the smallest possible particle of a chemical element that retains its chemical properties.
Molecule	A molecule is the smallest particle of a pure chemical substance that still retains its chemical composition and properties.
Calorie	Calorie refers to the amount of energy that raises the temperature of 1 g of water by 1°C.
Soil	Soil is material capable of supporting plant life. Soil forms through a variety of soil formation processes, and includes weathered rock "parent material" combined with dead and living organic matter and air.
Water vapor	Water vapor refers to water molecules in the gaseous state. Also is one state of the water cycle.
Salinity	Salinity refers to the total amount of salts dissolved in seawater. It is generally expressed in parts per thousand.
Element	A chemical element, often called simply element, is a chemical substance that cannot be divided or changed into other chemical substances by any ordinary chemical technique. An element is a class of substances that contain the same number of protons in all its atoms.
Residence time	Residence time refers to the ratio of the size of a compartment to the flux through it, expressed in units of time; thus, the average time spent by energy or a substance in the compartment.
Constant proportions	A principle that states that every pure substance always contains the same elements combined in the same proportions by weight is called the rule of constant proportions.
Solute	Substance that is dissolved in a solvent, forming a solution is referred to as a solute.
Potassium	Potassium is a chemical element in the periodic table. It has the symbol K (L. kalium) and atomic number 19. Potassium is a soft silvery-white metallic alkali metal that occurs naturally bound to other elements in seawater and many minerals.
Magnesium	Magnesium is the chemical element in the periodic table that has the symbol Mg and atomic number 12 and an atomic mass of 24.31.
Calcium	Calcium is the chemical element in the periodic table that has the symbol Ca and atomic number 20. Calcium is a soft grey alkaline earth metal that is used as a reducing agent in the extraction of thorium, zirconium and uranium. Calcium is also the fifth most abundant element in the Earth's crust.
Sulfur	Sulfur is the chemical element in the periodic table that has the symbol S and atomic number 16. It is an abundant, tasteless, odorless, multivalent non-metal. Sulfur, in its native form, is a yellow crystaline solid. In nature, it can be found as the pure element or as sulfide and sulfate minerals.
Sodium	Sodium is the chemical element in the periodic table that has the symbol Na (Natrium in Latin) and atomic number 11. Sodium is a soft, waxy, silvery reactive metal belonging to the alkali metals that is abundant in natural compounds (especially halite). It is highly reactive.
Residue	A residue refers to a portion of a larger molecule, a specific monomer of a polysaccharide,

protein or nucleic acid.

Trace element Trace element refers to an element that is essential for the survival of an organism but only in minute quantities.

Phosphorus Phosphorus is the chemical element in the periodic table that has the symbol P and atomic number 15.

Nitrogen A colorless and tasteless and mostly inert diatomic non-metal gas that is an essential constituent of proteins is nitrogen.

Iron Iron is essential to all organisms, except for a few bacteria. It is mostly stably incorporated in the inside of metalloproteins, because in exposed or in free form it causes production of free radicals that are generally toxic to cells.

Refraction Refraction is the change in direction of a wave due to a change in the velocity of the wave. It's most often demonstrated in optics, but may happen for any wave that interacts with a medium, such as sound waves.

Runoff Water entering rivers, lakes, reservoirs, or the ocean from land surfaces is runoff.

Photosynthesis Photosynthesis is a biochemical process in which plants, algae, and some bacteria harness the energy of light to produce food. Ultimately, nearly all living things depend on energy produced from photosynthesis for their nourishment, making it vital to life on Earth.

Respiration Respiration is the process by which an organism obtains energy by reacting oxygen with glucose to give water, carbon dioxide and ATP (energy). Respiration takes place on a cellular level in the mitochondria of the cells and provide the cells with energy.

Atmosphere Earth's atmosphere is a layer of gases surrounding the planet Earth and retained by the Earth's gravity. It contains roughly 78% nitrogen and 21% oxygen, with trace amounts of other gases.

Oxidation Oxidation refers to the loss of electrons from a substance involved in a redox reaction; always accompanies reduction.

Surface water Includes all bodies of water, lakes, rivers, ponds, and so on that are on the surface of the earth, in contrast to groundwater, which lies below the surface is surface water.

Oxygen minimum layer Oxygen minimum layer refers to a layer of water at a depth of approximately 500 m where the concentration of dissolved oxygen is very low.

Water column The vertical column of seawater that extends from the surface to the bottom is called water column.

Dissolved oxygen Oxygen gas molecules dissolved in water. Fish and other aquatic organisms are dependent on dissolved oxygen for respiration. Therefore concentration of dissolved oxygen is a measure of water quality.

Carbohydrate Carbohydrate is a chemical compound that contains oxygen, hydrogen, and carbon atoms. They consist of monosaccharide sugars of varying chain lengths and that have the general chemical formula $C_n(H_2O)_n$ or are derivatives of such.

Glucose Glucose, a simple monosaccharide sugar, is one of the most important carbohydrates and is used as a source of energy in animals and plants. Glucose is one of the main products of photosynthesis and starts respiration.

Decomposition Decomposition refers to the reduction of the body of a formerly living organism into simpler forms of matter.

Fjord A fjord is a narrow inlet of the sea between cliffs or steep slopes, which results from marine inundation of a glaciated valley. Typical characteristics of a fjord include: a narrow

	inlet, a bottom glacially eroded significantly below sea level, steep-sided walls which continue to descend below the sea surface, greater depths in the upper and middle reaches than on the seaward side, and communication with the open sea.
Predator	A predator is an animal or other organism that hunts and kills other organisms for food in an act called predation.
Protein	A protein is a complex, high-molecular-weight organic compound that consists of amino acids joined by peptide bonds. They are essential to the structure and function of all living cells and viruses. Many are enzymes or subunits of enzymes.
Pigment	Pigment is any material resulting in color in plant or animal cells which is the result of selective absorption.
Coral	Group of cnidarians having a calcium carbonate skeleton that participate in the formation of reefs is coral.
Coriolis effect	The deflection of large bodies of water or air to the right in the Northern Hemisphere and to the left in the Southern Hemisphere is referred to as the coriolis effect.
Trade winds	The trade winds are a pattern of wind found in bands around the Earth's equatorial region. The trade winds are the prevailing winds in the tropics, blowing from the high-pressure area in the horse latitudes towards the low-pressure area around the equator.
Westerlies	The westerlies are the prevailing winds in the middle latitudes between 30 and 60 degrees latitude, blowing from the high pressure area in the horse latitudes towards the poles. The winds are predominantly from the southwest in the Northern Hemisphere and from the northwest in the Southern Hemisphere.
Gyre	Gyre refers to a roughly circular pattern of ocean currents, formed because continents interrupt the currents' flow; rotates clockwise in the Northern Hemisphere and counterclockwise in the Southern Hemisphere.
Upwelling	Upward movement of deep, nutrient-rich water along coasts is referred to as upwelling.
Surface layer	The upper layer of water that is mixed by wind, waves, and currents is the surface layer .
Lens	The lens or crystalline lens is a transparent, biconvex structure in the eye that, along with the cornea, helps to refract light to focus on the retina. Its function is thus similar to a man-made optical lens.
Coral reef	A coral reef is a type of biotic reef that develops in tropical waters by the growth of coralline algae, hermatypic corals, and other marine organisms.
Tsunami	Long, fast waves produced by earthquakes and other seismic disturbances of the sea floor is called tsunami.
Gradient	Gradient refers to a difference in concentration, pressure, or electrical charge between two regions.
Thermocline	The thermocline is a layer within a body of water where the temperature changes rapidly with depth. Because water is not perfectly transparent, almost all sunlight is absorbed in the surface layer, which heats up.
Stratification	Stratification is the process of pretreating seeds to simulate natural conditions that a seed must endure before germination. Many seed species have what is called an embryonic dormancy and generally speaking will not sprout until this dormancy is broken.
Water mass	A body of water that can be identified by its temperature and salinity is water mass.
Deep sea	The deep sea refers to those areas of oceans to which little or no light penetrates.

Ocean current	An ocean current is any more or less permanent or continuous, directed movement of ocean water that flows in one of the Earth's oceans.
Fetch	The span of the sea surface over which the wind blows to form wind-driven waves is referred to as fetch.
Wavelength	The distance between crests of adjacent waves, such as those of the electromagnetic spectrum is wavelength.
Orbital	Orbital refers to a volume of space in which a given electron occurs 90% of the time. It is the quantum state of the individual electrons in the electron cloud around a single atom.
Erosion	Erosion is the displacement of solids (soil, mud, rock, and other particles) by the agents of wind, water, ice, movement in response to gravity, or living organisms.
Barrier island	A long and narrow island that is built by waves along the coast is called barrier island.
Tide	The periodic, rhythmic rise and fall of the sea surface caused by changes in gravitational forces external to the Earth is referred to as the tide.
Vector	A vector is an organism that does not cause disease itself but which spreads infection by conveying pathogens from one host to another.
Diurnal	Diurnal is a term that refers to being active in the daytime.
Semidiurnal tide	Semidiurnal tide is a tide having two high waters and two low waters each lunar day, with little or no diurnal inequality. This applies equally to solar tides and to atmospheric tides. Worldwide, they are the most important because the global ocean is near to resonance at the period of semidiurnal gravitational forcing.
Diurnal tide	A tidal pattern with a high and a low tide each day is called a diurnal tide.
Tidal range	The difference in water level between successive high and low tides is the tidal range.
Neap tides	Neap tides refers to the tides with a small tidal range. Neap tides occur when the gravitational forces of the Moon and the Sun are perpendicular to one another and occur during quarter moons.
Salt marsh	Salt marsh refers to a grassy area that extends along the shores of estuaries and sheltered coasts in temperate and subpolar regions.
Estuary	An estuary is a semi-enclosed coastal body of water which has a free connection with the open sea and within which sea water mixes with fresh water. The key feature of an estuary is that it is a mixing place for sea water and a stream or river to supply fresh water.
Stem	Stem refers to that part of a plant's shoot system that supports the leaves and reproductive structures.
Lagoon	A lagoon is a body of comparatively shallow salt water separated from the deeper sea by a shallow or exposed sandbank, coral reef, or similar feature. Thus, the enclosed body of water behind a barrier reef or barrier islands or enclosed by an atoll reef is called a lagoon.
Downstream	Downstream refers to a relative position in DNA or RNA. Relative to the position on the strand, downstream is the region towards the 3' end of the strand.
Homogeneous	Homogeneous refers both to animals and plants, of having a resemblance in structure, due to descent from a common progenitor with subsequent modification.
Particulate matter	In air pollution, solid particles and liquid droplets, as opposed to material uniformly dispersed among air molecules is particulate matter.
Watershed	The total land area that drains directly or indirectly into a particular stream or river. The watershed is generally named for the stream or river into which it drains.

Paleoecology	Paleoecology uses the data of fossils or subfossils to reconstruct the ecosystems of the past. It includes the study of fossil organisms in terms of their life cycle, their interactions, their natural environment, their manner of death and burial.
Ecology	Ecology is the scientific study of the distribution and abundance of living organisms and how these properties are affected by interactions between the organisms and their environment.
Population	Group of organisms of the same species occupying a certain area and sharing a common gene pool is referred to as population.
Ecosystem	In general terms an ecosystem can be thought of as an assemblage of organisms (plant, animal and other living organisms living together with their environment, functioning as a loose unit. That is, a dynamic and complex whole, interacting as an "ecological unit".
Systematics	In biology, systematics is the study of the diversity of organism characteristics, and especially how they relate evolutionarily. The term "systematics" may or may not overlap with "taxonomy" which concerns itself with scientific classification of species and other taxa.
Desert	A desert is a landscape form or region that receives little precipitation - less than 250 mm (10 in) per year. It is a biome characterized by organisms adapted to sparse rainfall and rapid evaporation.
Evolution	In biology, evolution is the process by which novel traits arise in populations and are passed on from generation to generation. Its action over large stretches of time explains the origin of new species and ultimately the vast diversity of the biological world.
Species	Group of similarly constructed organisms capable of interbreeding and producing fertile offspring is a species.

Ecology	Ecology is the scientific study of the distribution and abundance of living organisms and how these properties are affected by interactions between the organisms and their environment.
Biological interaction	A biological interaction results from the fact that organisms in an ecosystem interact with each other, in the natural world, no organism is an autonomous entity isolated from its surroundings.
Parasite	A parasite is an organism that spends a significant portion of its life in or on the living tissue of a host organism and which causes harm to the host without immediately killing it. They also commonly show highly specialized adaptations allowing them to exploit host resources.
Predator	A predator is an animal or other organism that hunts and kills other organisms for food in an act called predation.
Species	Group of similarly constructed organisms capable of interbreeding and producing fertile offspring is a species.
Host	Host is an organism that harbors a parasite, mutual partner, or commensal partner; or a cell infected by a virus.
Abiotic	Abiotic refers to the nonliving portion of an ecosystem. It includes soil, rock, water, and the atmosphere.
Distribution	Distribution in pharmacology is a branch of pharmacokinetics describing reversible transfer of drug from one location to another within the body.
Reproduction	Biological reproduction is the biological process by which new individual organisms are produced. Reproduction is a fundamental feature of all known life; each individual organism exists as the result of reproduction by an antecedent.
Protozoa	Protozoa are single-celled eukaryotes (organisms with nuclei) that show some characteristics usually associated with animals, most notably mobility and heterotrophy. They are often grouped in the kingdom Protista together with the plant-like algae.
Population	Group of organisms of the same species occupying a certain area and sharing a common gene pool is referred to as population.
Ecosystem	In general terms an ecosystem can be thought of as an assemblage of organisms (plant, animal and other living organisms living together with their environment, functioning as a loose unit. That is, a dynamic and complex whole, interacting as an "ecological unit".
Biosphere	The biosphere is that part of a planet's outer shell — including air, land, surface rocks and water — within which life occurs, and which biotic processes in turn alter or transform.
Extinction	In biology and ecology, extinction is the ceasing of existence of a species or group of taxa. The moment of extinction is generally considered to be the death of the last individual of that species.The death of all members of a species is extinction.
Biogeography	Biogeography is the science which deals with patterns of species distribution and the processes that result in such patterns.
Climate	Weather condition of an area including especially prevailing temperature and average daily/yearly rainfall over a long period of time is called climate.
Starfish	Starfish are marine invertebrates belonging to phylum Echinodermata, class Asteroidea. They exhibit a superficially radial symmetry, typically with five or more "arms" which radiate from an indistinct disk. They do not have movable skeletons, but instead possess a hydraulic water vascular system.
Barnacle	A barnacle is a type of arthropod belonging to infraclass Cirripedia in the subphylum

	Crustacea and is hence distantly related to crabs and lobsters. They live attached to surfaces and are typically enclosed by heavy calcareous plates are referred to as barnacles.
Snail	The name snail applies to most members of the molluscan class Gastropoda that have coiled shells. Other gastropods, which lack a conspicuous shell, are commonly called slugs, and are scattered throughout groups that primarily include snails.
Habitat	Habitat refers to a place where an organism lives; an environmental situation in which an organism lives.
Estuary	An estuary is a semi-enclosed coastal body of water which has a free connection with the open sea and within which sea water mixes with fresh water. The key feature of an estuary is that it is a mixing place for sea water and a stream or river to supply fresh water.
Carbon dioxide	Carbon dioxide is an atmospheric gas comprized of one carbon and two oxygen atoms. A very widely known chemical compound, it is frequently called by its formula CO_2. In its solid state, it is commonly known as dry ice.
Atmosphere	Earth's atmosphere is a layer of gases surrounding the planet Earth and retained by the Earth's gravity. It contains roughly 78% nitrogen and 21% oxygen, with trace amounts of other gases.
Photosynthesis	Photosynthesis is a biochemical process in which plants, algae, and some bacteria harness the energy of light to produce food. Ultimately, nearly all living things depend on energy produced from photosynthesis for their nourishment, making it vital to life on Earth.
Phytoplankton	Phytoplankton refers to the autotrophic component of the plankton that drifts in the water column. Phytoplankton obtain energy through a process called photosynthesis, and so must live in the well-lit surface layer (termed the euphotic zone) of an ocean, sea, or lake.
Biotic	Biotic refers to life or living organisms.
Salinity	Salinity refers to the total amount of salts dissolved in seawater. It is generally expressed in parts per thousand.
Territory	In ethology, sociobiology and behavioral ecology, the term territory refers to any geographical area that an animal of a particular species consistently defends against conspecifics (and, occasionally, animals of other species).
Predation	Interaction in which one organism uses another, called the prey, as a food source is referred to as predation.
Gastropod	A member of the largest group of mollusks, including snails and slugs is a gastropod.
Crab	The term crab is often applied to several different groups of short (nose to tail) decapod crustaceans with thick exoskeletons, but only members of the Brachyura are true.
Mussel	A mussel is a bivalve mollusk that can be found in lakes, rivers, creeks, intertidal areas, and throughout the ocean. The saltwater mussels (family Mytilidae) and freshwater mussels (family Unionidae) are not thought to be closely related and are grouped in different subclasses, despite considerable similarities in appearance.
Invertebrate	Invertebrate is a term coined by Jean-Baptiste Lamarck to describe any animal without a spinal column. It therefore includes all animals except vertebrates (fish, reptiles, amphibians, birds and mammals).
Imprinting	Learning that is limited to a specific critical period in an animal's life and that is generally irreversible is called imprinting.
Bivalve	Bivalve refers to a member of a group of mollusks that includes clams, mussels, scallops, and oysters. They typically have two-part shells, with both parts being more or less symmetrical.

Mollusk	The mollusk is the large and diverse phylum Mollusca, which includes a variety of familiar creatures well-known for their decorative shells or as seafood. These range from tiny snails, clams, and abalone to the octopus, cuttlefish and squid (which are considered the most intelligent invertebrates).
Siphon	The tube-like extension through which water flows in and out of the mantle cavity in bivalves and cephalopods, and in tunicates is called siphon.
Organ	Organ refers to a structure consisting of several tissues adapted as a group to perform specific functions.
Natural selection	Natural selection is the process by which biological individuals that are endowed with favorable or deleterious traits end up reproducing more or less than other individuals that do not possess such traits.
Trait	In biology, a trait or character is a genetically inherited feature of an organism: Eye color is a character; brown eyes and blue eyes is a trait.
Joule	The joule is a derived unit defined as the work done or energy required, to exert a force of one newton for a distance of one metre, so the same quantity may be referred to as a newton metre or newton-metre with the symbol N·m.
Adaptation	A biological adaptation is an anatomical structure, physiological process or behavioral trait of an organism that has evolved over a period of time by the process of natural selection such that it increases the expected long-term reproductive success of the organism.
Chromatophore	A pigment-bearing cell that found in amphibians, fish, reptiles and cephalopods that expands or contracts to change the color of the organism is called chromatophore.
Morphology	The scientific study of organic form, including both its development and function is morphology.
Gene	Gene refers to a discrete unit of hereditary information consisting of a specific nucleotide sequence in DNA . Most of the genes of a eukaryote are located in its chromosomal DNA; a few are carried by the DNA of mitochondria and chloroplasts.
Crypsis	Crypsis is a phenomena where an organism's appearance allows it to blend well into its environment.
Crustacean	Member of a group of marine arthropods that contains among others shrimps, crabs, crayfish, and lobsters is called crustacean.
Cephalopod	Cephalopod refers to a member of a group of mollusks that includes squids and octopuses, characterized by bilateral body symmetry.
Cell	The cell is the structural and functional unit of all living organisms, and is sometimes called the "building block of life."
Dorsal surface	The dorsal surface is arched from above downward, and is subdivided into two unequal parts by the spine; the portion above the spine is called the supraspinatous fossa, and that below it the infraspinous fossa.
Carapace	Carapace refers to the shield-like structure that covers the anterior portion of some crustaceans. The shell of sea turtles.
Seaweed	Phycologists consider seaweed to refer any of a large number of marine benthic algae that are multicellular, macrothallic (large-bodied), and thus differentiated from most algae that tend to be microscopic in size.
Sponge	An invertebrates that consist of a complex aggregation of cells, including collar cells, and has a skeleton of fibers and/or spicules is a sponge. They are primitive, sessile, mostly

marine, waterdwelling filter feeders that pump water through their matrix to filter out particulates of food matter.

Ventral — The surface or side of the body normally oriented upwards, away from the pull of gravity, is the dorsal side; the opposite side, typically the one closest to the ground when walking on all legs, swimming or flying, is the ventral side.

Bryozoan — Colonial animals that build stony skeletons of calcium carbonate and superficially resemble coral is referred to as bryozoan.

Diatom — Diatom refers to a unicellular photosynthetic alga with a unique, glassy cell wall containing silica.

Water column — The vertical column of seawater that extends from the surface to the bottom is called water column.

Substratum — Substratum is the underlying layer, or base to which a lichen is fixed.

Sea cucumber — The sea cucumber is an echinoderm of the class Holothuroidea, with an elongated body and leathery skin. Mostly found on the sea floor.

Muscle — Muscle is a contractile form of tissue. It is one of the four major tissue types, the other three being epithelium, connective tissue and nervous tissue. Muscle contraction is used to move parts of the body, as well as to move substances within the body.

Skeleton — In biology, the skeleton or skeletal system is the biological system providing physical support in living organisms.

Spine — A spine is a rigid, pointed surface protuberance or needle-like structure on an animal, shell, or plant, presumably serving as a defense against attack by predators.

Calcium carbonate — A mineral that is the major component of the shell, skeleton, and other parts of many organisms are called calcium carbonate.

Thallus — Thallus is an undifferentiated vegetative tissue of some non-mobile organisms, which were previously known as the thallophytes.

Coral reef — A coral reef is a type of biotic reef that develops in tropical waters by the growth of coralline algae, hermatypic corals, and other marine organisms.

Diadema antillarum — Diadema antillarum is a sea urchin of the genus diadema with long black spines, it was once the most abundant and important herbivore on the coral reefs of the western Atlantic and Caribbean basin. In this role it was the main grazer that maintained the balance between algae growth and destruction.

Sessile — In botany, sessile means without a stalk, as in flowers or leaves that grow directly from the stem. In zoology, sessile organisms are those which are not able to move about. They are usually permanently attached to a solid substrate of some kind, such as a rock, or the hull of a ship in the case of barnacles.

Colony — Colony refers to a cluster or assemblage of microorganisms growing on the surface of an agar culture medium. A colony also refers to several individual organisms of the same species living closely together, usually for mutual benefit, such as stronger defences, the ability to attack bigger prey, etc.

Nematocyst — In cnidarians, a capsule that contains a threadlike fiber whose release aids in the capture of prey is referred to as nematocyst.

Stolon — A stolon is an aerial shoot from a plant with the ability to produce adventitious roots and new offshoots of the same plant.

Fecundity — Fecundity refers to the ability to reproduce. In biology and demography, fecundity is the

	potential reproductive capacity of an organism or population, measured by the number of gametes (eggs), seed set or asexual propagules.
Somatic	The term somatic refers to the body. It also refers to the part of the nervous system that controls voluntary movement and senzation and judges relative effort and weight, called proprioception.
Carnivore	An animal that eats a diet consisting solely of meat is referred to as a carnivore.
Secondary compound	Secondary compound refers to a compound synthesized by a plant that is not needed for basic cellular metabolism. Typically has an antiherbivore or antiparasite function.
Secretion	Secretion is the process of segregating, elaborating, and releasing chemicals from a cell, or a secreted chemical substance or amount of substance.
Acid	An acid is a water-soluble, sour-tasting chemical compound that when dissolved in water, gives a solution with a pH of less than 7.
Stimulant	A stimulant is a drug which increases the activity of the sympathetic nervous system and produces a sense of euphoria or awakeness.
Metabolic processes	The total of all chemical reactions within an organism are the metabolic processes.
Organic compound	An organic compound is any member of a large class of chemical compounds whose molecules contain carbon, with the exception of carbides, carbonates, carbon oxides and gases containing carbon.
Tunicate	One of a group of invertebrate chordates in the subphylum of saclike filter feeders with input and output siphons is referred to as a tunicate.
Sulfuric acid	Sulfuric acid refers to the major constituent of acid precipitation. Formed as a result of sulfur dioxide emissions reacting with water vapor in the atmosphere.
Vacuole	A vacuole is a large membrane-bound compartment within some eukaryotic cells where they serve a variety of different functions: capturing food materials or unwanted structural debris surrounding the cell, sequestering materials that might be toxic to the cell, maintaining fluid balance (called turgor) within the cell.
Tentacle	An elongate, extensible projection of the body of cnidarians and cephalopod mollusks that may be used for grasping, stinging, and immobilizing prey, and for locomotion is called tentacle.
Mantle	The mantle is an organ found in mollusks. It is the dorsal body wall covering the main body, or visceral mass. The epidermis of this organ secretes calcium carbonate to create a shell.
Microhabitat	A microhabitat is the immediate surroundings and other physical factors of an individual plant or animal within its habitat.
Gradient	Gradient refers to a difference in concentration, pressure, or electrical charge between two regions.
Amphipods	Amphipods include over 7000 described species of small, shrimp-like crustaceans. Most amphipods are marine, although a few live in freshwater or are terrestrial. One species common to mid-Atlantic and estuarine waters is Jassa falcata.
Herbivore	A herbivore is an animal that is adapted to eat primarily plant matter
Commensalism	Symbiotic relationship in which one species is benefited and the other is neither harmed nor benefited is commensalism.
Commensal	Living on or within another organism without injuring or benefiting the other organism is referred to as commensal.

Larva	A free-living, sexually immature form in some animal life cycles that may differ from the adult in morphology, nutrition, and habitat is called larva.
Mucus	Mucus is a slippery secretion of the lining of various membranes in the body (mucous membranes). Mucus aids in the protection of the lungs by trapping foreign particles that enter the nose during normal breathing. Additionally, it prevents tissues from drying out.
Detritus	Detritus is organic waste material from decomposing dead plants or animals.
Mammal	Homeothermic vertebrate characterized especially by the presence of hair and mammary glands is a mammal.
Dorsal	In anatomy, the dorsal is the side in which the backbone is located. This is usually the top of an animal, although in humans it refers to the back.
Mutualism	Mutualism is an interaction between two species in which both species derive benefit.
Trade-off	Trade-off refers to a balancing of factors, all of which are not attainable at the same time, especially traits that tend to increase the evolutionary fitness or performance of the individual.
Ectoparasite	A parasite that lives on or attached to the host's surface is referred to as ectoparasite.
Parasitism	A symbiotic relationship in which the parasite, a type of predator, lives within or on the surface of a host, from which it derives its food is a parasitism.
Gill	An extension of the body surface of an aquatic animal, specialized for gas exchange and/or suspension feeding is called a gill.
Endoparasite	A parasite that lives within the tissues or bloodstream of its host is referred to as endoparasite.
Absorption	Absorption is a physical or chemical phenomenon or a process in which atoms, molecules, or ions enter some bulk phase - gas, liquid or solid material. In nutrition, amino acids are broken down through digestion, which begins in the stomach.
Tissue	Group of similar cells which perform a common function is called tissue.
Gonad	Gonad refers to a sex organ in an animal; an ovary or a testis. It is the organ that makes gametes.
Dispersal	Biological dispersal refers to those processes by which a species maintains or expands the distribution of a population. Dispersal implies movement—movement away from an existing population or away from the parent organisms.
Isopod	Isopod is one of the most diverse orders of Crustaceans, with many species living in all environments, and are common in shallow marine waters. Unlike most crustaceans, isopods are successful on land (suborder Oniscidea, woodlice), although their greatest diversity remains in the deep sea (suborder Asellota).
Flatworm	The flatworm is a phylum of relatively simple soft-bodied invertebrate animals. With about 25,000 known species they are the largest phylum of acoelomates. They are found in marine, freshwater, and even damp terrestrial environments.
Phylum	Phylum is the second highest taxonomic classification for the kingdom Animalia (animals), between kingdom level and class level.
Intermediate host	In parasitology, an intermediate host is an organism within which a parasite does not sexually reproduce. It may grow or reproduce asexually, but there is no exchange of genetic material between individuals.
Population	The number of organisms of a species per unit area is called population density.

density	
Spawning	A method of external fertilization in which male and female parents shed gametes into the water, and sperm must swim through the water to reach the eggs is referred to as spawning.
Biomass	Biomass is organic non-fossil material, collectively. In other words, biomass comprises the mass of all biological organisms, dead or alive, excluding biological mass that has been transformed by geological processes into substances such as coal or petroleum.
Geographic barriers	Geographic features that keep different portions of a species from exchanging genes are called geographic barriers.
Conception	Conception is fusion of gametes to form a new organism. In animals, the process involves a sperm fusing with an ovum, which eventually leads to the development of an embryo.
Geographic range	The geographic area in which a population lives is a geographic range.
Stem	Stem refers to that part of a plant's shoot system that supports the leaves and reproductive structures.
Growth ring	A growth ring can be seen in a horizontal cross section cut through the trunk of a tree. Visible rings result from the change in growth speed through the seasons of the year, thus one ring usually marks the passage of one year in the life of the tree.
Coral	Group of cnidarians having a calcium carbonate skeleton that participate in the formation of reefs is coral.
Survivorship curve	Survivorship curve refers to a plot of the number of members of a cohort that are still alive at each age; one way to represent age-specific mortality.
Generation time	Generation time refers to the time between the birth of a parent and the birth of its offspring. The time interval between successive cell divisions.
Egg	An egg is the zygote, resulting from fertilization of the ovum. It nourishes and protects the embryo.
Ocean current	An ocean current is any more or less permanent or continuous, directed movement of ocean water that flows in one of the Earth's oceans.
Catastrophe	An unpredictable event that has a strong effect on a population is referred to as a catastrophe.
Carrying capacity	Carrying capacity is the measure of habitat to indefinitely sustain a population at a particular density.
Metapopulation	A metapopulation consists of a group of spatially separated populations of the same species which interact at some level.
Phenotypic plasticity	The ability of a genotype to change its phenotype in response to changes in the environment is called phenotypic plasticity.
Homogeneous	Homogeneous refers both to animals and plants, of having a resemblance in structure, due to descent from a common progenitor with subsequent modification.
Substrate	A substrate is a molecule which is acted upon by an enzyme. Each enzyme recognizes only the specific substrate of the reaction it catalyzes. A surface in or on which an organism lives.
Mangrove	Mangrove are woody trees or shrubs that grow in coastal habitats or mangal (Hogarth, 1999), for which the term mangrove swamp also would apply.
Leaf	In botany, a leaf is an above-ground plant organ specialized for photosynthesis. For this purpose, a leaf is typically flat (laminar) and thin, to expose the chloroplast containing cells (chlorenchyma tissue) to light over a broad area, and to allow light to penetrate fully

	into the tissues.
Mudflat	Mudflat is a coastal wetland that forms when mud is deposited by the tides or rivers, they are found in sheltered areas such as bays, bayous, lagoons, and estuaries.
Uniform distribution	The distribution characteristic of a population with a relatively regular spacing of individuals, commonly as a result of territorial behavior is called uniform distribution.
Taxonomic classification	Taxonomic classification is the act of placing an object or concept into a set or sets of categories (such as a taxonomy or a subject index), based on the properties of the object or concept.
Genus	In biology, a genus is a taxonomic grouping. That is, in the classification of living organisms, a genus is considered to be distinct from other such genera. A genus has one or more species: if it has more than one species these are likely to be morphologically more similar than species belonging to different genera.
Subspecies	Regional groups within a species that are significantly different structurally, physiologically, or behaviorally, yet are capable of exchanging genes by interbreeding is a subspecies.
Derived character	A derived character is homologous feature that has changed from a primitive condition and is unique to an evolutionary lineage; feature found in members of a lineage but not found in ancestors of the lineage.
Horseshoe crab	Horseshoe crab refers to a bottom-dwelling arthropod that is more closely related to spiders than crabs, that belongs to the phylum Arthropoda.
Segmentation	Segmentation in biology refers to the division of some metazoan bodies and plant body plans into a series of semi-repetitive segments, and the question of the benefits and costs of doing so.
Arthropod	Arthropod refers to member of a phylum of invertebrates that contains among other groups crustaceans and insects that have an exoskeleton and jointed appendages.
Cuticle	Cuticle in animals, a tough, nonliving outer layer of the skin. In plants, a waxy coating on the surface of stems and leaves that helps retain water.
Insect	An arthropod that usually has three body segments , three pairs of legs, and one or two pairs of wings is called an insect. They are the largest and (on land) most widely-distributed taxon within the phylum Arthropoda. They comprise the most diverse group of animals on the earth, with around 925,000 species described
Fossil record	The chronicle of evolution over millions of years of geologic time engraved in the order in which fossils appear in rock strata is called a fossil record.
Biochemistry	Biochemistry studies how complex chemical reactions give rise to life. It is a hybrid branch of chemistry which specialises in the chemical processes in living organisms.
Polymorphism	The presence in a population of more than one allele of a gene at a frequency greater than that of newly arising mutations is referred to as polymorphism.
Genotype	The genotype is the specific genetic makeup (the specific genome) of an individual, usually in the form of DNA. It codes for the phenotype of that individual.
Eye	An eye is an organ that detects light. Different kinds of light-sensitive organs are found in a variety of creatures.
Phenotype	The phenotype of an individual organism is either its total physical appearance and constitution or a specific manifestation of a trait, such as size or eye color, that varies between individuals. It is determined to some extent by genotype.

Downstream	Downstream refers to a relative position in DNA or RNA. Relative to the position on the strand, downstream is the region towards the 3' end of the strand.
Node	In botany, a node is the place on a stem where a leaf is attached.
Circulatory system	The circulatory system or cardiovascular system is the organ system which circulates blood around the body of most animals.
Spicule	Spicule refers to a subunit of the endoskeleton of sponges that is made of silica or calcium carbonate.
Coelom	A fluid filled body cavity with a complete lining called peritoneum derived from mesoderm is called a coelom. Organs formed inside a coelom can freely move, grow, and develop independently of the body wall while fluid cushions and protects them from shocks.
Gland	A gland is an organ in an animal's body that synthesizes a substance for release such as hormones, often into the bloodstream or into cavities inside the body or its outer surface.
Allele	An allele is any one of a number of viable DNA codings of the same gene (sometimes the term refers to a non-gene sequence) occupying a given locus (position) on a chromosome.
Heterozygote	Heterozygote refers to a diploid or polypoid individual carrying two different alleles of a gene on its two homologous chromosomes.
Chromosome	A chromosome is, minimally, a very long, continuous piece of DNA, which contains many genes, regulatory elements and other intervening nucleotide sequences.
Protein	A protein is a complex, high-molecular-weight organic compound that consists of amino acids joined by peptide bonds. They are essential to the structure and function of all living cells and viruses. Many are enzymes or subunits of enzymes.
Oxygen	Oxygen is a chemical element in the periodic table. It has the symbol O and atomic number 8. Oxygen is the second most common element on Earth, composing around 46% of the mass of Earth's crust and 28% of the mass of Earth as a whole, and is the third most common element in the universe.
Enzyme	An enzyme is a protein that catalyzes, or speeds up, a chemical reaction. They are essential to sustain life because most chemical reactions in biological cells would occur too slowly, or would lead to different products, without them.
DNA sequence	A DNA sequence is a succession of letters representing the primary structure of a real or hypothetical DNA molecule or strand, The possible letters are A, C, G, and T, representing the four nucleotide subunits of a DNA strand (adenine, cytosine, guanine, thymine).
Sex-determining chromosome	The chromosomes X and Y that are primarily responsible for determining if an individual will develop as a male or female is a sex-determining chromosome.
Dominance	Dominance is the state of having high social status relative to other individuals, who react submissively to dominant individuals.
Genetics	Genetics is the science of genes, heredity, and the variation of organisms.
Cline	Cline refers to a gradation in an inherited trait along a geographic continuum; variation in a population's phenotypic features that parallels an environmental gradient.
Dehydrogenase	An enzyme that catalyzes a chemical reaction during which one or more hydrogen atoms are removed from a molecule is dehydrogenase.
Isthmus	An isthmus is a narrow strip of land that is bordered on two sides by water and connects two larger land masses. It is the inverse of a strait (which lies between two land masses and connects two larger bodies of water).

Evolution	In biology, evolution is the process by which novel traits arise in populations and are passed on from generation to generation. Its action over large stretches of time explains the origin of new species and ultimately the vast diversity of the biological world.
Sediment	Sediment is any particulate matter that can be transported by fluid flow and which eventually is deposited as a layer of solid particles on the bed or bottom of a body of water or other liquid.
Deep ocean	The deep ocean is the lowest layer in an ocean, existing below the thermocline. The deep ocean is not well mixed, consists of horizontal layers of equal density, and is often as cold as -1 to 4 degrees Celsius. Ninety percent of the total volume of Earth's oceans is found in the deep ocean.
Biota	Biota is the plant and animal life of a region or area or geological period. The biota of the Earth is known as the biosphere.
Temperate	In virology, temperate refers to the life cycle a phage is able to perform. The phage usually integrates its genome into its host chromosome, becoming a prophage.
Boreal	Boreal is usually applied to ecosystems localized to subarctic (Northern hemisphere) and subantarctic (Southern hemisphere) zones, although Austral is also used for the latter. A boreal forest, also known as the taiga, is the set of forest ecosystems that can survive in northern, specifically subarctic, regions.
Speciation	Speciation refers to the evolutionary process by which new biological species arise. All forms of speciation have actually taken place over the course of evolution, though it still remains a subject of debate as to the relative importance of each mechanism in driving biodiversity.
Brachiopod	A brachiopod is an animala that superficially resembles clams and other bivalve mollusks, but the two halves of the brachiopod shell are dorsal and ventral to the animal rather than lateral, as in clams.
Dissolved oxygen	Oxygen gas molecules dissolved in water. Fish and other aquatic organisms are dependent on dissolved oxygen for respiration. Therefore concentration of dissolved oxygen is a measure of water quality.
Limiting factor	In biology, agricultural science, physiology, and ecology, a limiting factor is one that controls a process, such as organism growth or species population size or distribution. The availability of food, predation pressure, or availability of shelter are examples of factors that could be limiting for a species population in a specific area.
Interspecific competition	Competition between populations of two or more species for one or more of the same limited resources of food, sunlight, water, soil, nutrients, or space. Interspecific competition may inhibit population growth and help structure communities.
Renewable resource	A renewable resource is any natural resource that is depleted at a rate slower than the rate at which it regenerates.
Copepods	A group of small crustaceans that are important members of marine and freshwater plankton communities is copepods.
Copepod	The group of small crustaceans that is an important member of marine and freshwater plankton communities is copepod. They are very important food organisms for small fish, whales, and other crustaceans such as krill in the ocean and in fresh water.
Excretion	Excretion is the biological process by which an organism chemically separates waste products from its body. The waste products are then usually expelled from the body by elimination.
Ammonia	Ammonia is a compound of nitrogen and hydrogen with the formula NH_3. At standard

	temperature and pressure ammonia is a gas. It is toxic and corrosive to some materials, and has a characteristic pungent odor.
Standing stock	Standing stock refers to the total amount, or biomass, of an organism at a given time.
Bacteria	The domain that contains procaryotic cells with primarily diacyl glycerol diesters in their membranes and with bacterial rRNA. Bacteria also is a general term for organisms that are composed of procaryotic cells and are not multicellular.
Coralline algae	Red algae that deposit calcium carbonate in their thallus are called the coralline algae.
Dominant species	Dominant species refers to those species in a community that have the highest abundance or highest biomass. These species exert a powerful control over the occurrence and distribution of other species.
Zonation	The distribution of organisms in bands or regions corresponding to changes in ecological conditions along a continuum, for example, intertidal zonation and elevational zonation.
Limpet	Limpet is a marine mollusk in the family Acmaeidae with flattened, cone-shaped shells. They live throughout the intertidal zone, from the high zone (upper littoral) to the shallow subtidal on the rocky coasts of most oceans. They can be commonly found attached to rocks, looking like little disks or bumps on the rock surface.
Lichen	Lichen is a symbiotic organism in a mutualistic association between a fungus and an alga or between a fungus and a cyanobacterium.
Bacterium	Most bacterium are microscopic and unicellular, with a relatively simple cell structure lacking a cell nucleus, and organelles such as mitochondria and chloroplasts. They are the most abundant of all organisms. They are ubiquitous in soil, water, and as symbionts of other organisms.
Deep sea	The deep sea refers to those areas of oceans to which little or no light penetrates.
Plankton	Plankton are drifting organisms that inhabit the water column of oceans, seas, and bodies of fresh water. Plankton abundance and distribution are strongly dependent on factors such as ambient nutrients concentrations, the physical state of the water column, and the abundance of other plankton.
Controlled experiment	An experiment that includes at least two groups with one group not being manipulated is called a controlled experiment.
Zooplankton	Nonphotosynthetic protists that feed on other plankton and are abundant in marine and freshwater environments are zooplankton.
Growth rate	A measure of the change in population size per individual per unit of time is called growth rate.
Tide	The periodic, rhythmic rise and fall of the sea surface caused by changes in gravitational forces external to the Earth is referred to as the tide.
Zooid	One of the distinct individuals forming a colonial animal such as a bryozoan or hydrozoan is called zooid.
Sweeper tentacle	A sweeper tentacle is an elongated tentacle equipped with stinging nematocyst cells found mainly on large polyped stony corals and used for defence.
Niche	Niche refers to the relational position of a species or population in an ecosystem. More formally, the niche includes how a population responds to the abundance of its resources and enemies and how it affects those same factors .
Species diversity	Species diversity refers to a weighted representation of the species of organisms living in a region; large and common species are given greater weight than are small and rare ones.

Coexistence	Coexistence refers to occurrence of two or more species in the same habitat; usually applied to potentially competing species.
Nitrogen	A colorless and tasteless and mostly inert diatomic non-metal gas that is an essential constituent of proteins is nitrogen.
Competitive exclusion	Competitive exclusion refers to the elimination of one species by another as a result of competition.
Colonization	Colonization is the act where life forms move into a distant area where their kind is sparse or not yet existing at all and set up new settlements in the area. Colonization applies to all life forms in a sense though it is most often used in reference to insects and humans.
Limestone	Limestone is a sedimentary rock composed largely of the mineral calcite (calcium carbonate: $CaCO_3$). Limestone often contains variable amounts of silica in the form of chert or flint, as well as varying amounts of clay, silt and sand as disseminations, nodules, or layers within the rock.
Sea urchin	A sea urchin is a spiny sea creature of the class Echinoidea found in oceans all over the world. Their shell, which biologists call the test, is globular in shape, and covered with spines.
Intertidal zone	The part of a beach that lies between average high tide and average low tide is called the intertidal zone.
Migration	Migration occurs when living things move from one biome to another. In most cases organisms migrate to avoid local shortages of food, usually caused by winter. Animals may also migrate to a certain location to breed, as is the case with some fish.
Humidity	Humidity is the concentration of water vapor in the air. The concentration can be expressed as absolute humidity, specific humidity, or relative humidity.
Digestive tract	The digestive tract is the system of organs within multicellular animals which takes in food, digests it to extract energy and nutrients, and expels the remaining waste.
Tapeworm	A parasitic flatworm characterized by the absence of a digestive tract or a mouth is called a tapeworm.
Pathogen	A pathogen or infectious agent is a biological agent that causes disease or illness to its host.The term is most often used for agents that disrupt the normal physiology of a multicellular animal or plant.
Soil	Soil is material capable of supporting plant life. Soil forms through a variety of soil formation processes, and includes weathered rock "parent material" combined with dead and living organic matter and air.
Climax community	The term climax community is an ecological term for a community of plants and animals which is the result of succession, where a biological system, a community, or a soil has reached a steady state.
Positive feedback	Mechanism of homeostatic response in which the output intensifies and increases the likelihood of response, instead of countering it and canceling it is called positive feedback.
Grazing	Grazing refers to consumption of a portion of a plant's tissues, generallyapplied to grasses and other herbaceous vegetation; removing plant or algal cover from a substrate, generally applied to aquatic systems.
Green algae	Green algae refers to photosynthetic protists that include L.nicellular, colonial, and multicellular species with grass green chloroplasts; closely related to true plants.

Food web	A network of interconnecting food chains is referred to as a food web.
Molecule	A molecule is the smallest particle of a pure chemical substance that still retains its chemical composition and properties.
Primary producer	Primary producer refers to a photosynthetic or chemosynthetic organism that synthesizes complex organic molecules from simple inorganic ones.
Saprophyte	Saprophyte refers to an organism that obtains energy by the decomposition of dead organic material.
Chemical energy	Chemical energy refers to energy stored in the chemical bonds of molecules; a form of potential energy.
Primary productivity	Primary productivity is the amount of light energy converted to chemical energy in a given amount of time. It usually corresponds to the amount of photosynthesis that occurred.
Carbon	Carbon is a chemical element in the periodic table that has the symbol C and atomic number 6. An abundant nonmetallic, tetravalent element, carbon has several allotropic forms.
Sugar	A sugar is the simplest molecule that can be identified as a carbohydrate. These include monosaccharides and disaccharides, trisaccharides and the oligosaccharides. The term "glyco-" indicates the presence of a sugar in an otherwise non-carbohydrate substance.
Food chain	Food chain refers to a sequence of food transfers from one species to another within an ecosystem.
Secondary production	Secondary production is the biomass production resulting from the assimilation of organic matter produced by a primary consumer; production by organisms (mainly animals) which consume primary producers (mainly plants).
Trophic level	In ecology, the trophic level is the position that an organism occupies in a food chain - what it eats, and what eats it.
Keystone species	Species that are not usually abundant in a community yet exert strong control on community structure by the nature of their ecological roles or niches are keystone species.
Linkage	Linkage refers to the patterns of assortment of genes that are located on the same chromosome. Important because if the genes are located relatively far apart, crossing over is more likely to occur between them than if they are close together.
Functional group	Functional group is a submolecular structural motif, characterized by specific elemental composition and connectivity, that confer reactivity upon the molecule that contains it.
Evolutionary biology	Evolutionary biology is a subfield of biology concerned with the origin and descent of species, as well as their change over time, i.e. their evolution. One who studies evolutionary biology is known as an evolutionary biologist, or, less frequently, evolutionist.
Primer	A primer is a nucleic acid strand, or a related molecule that serves as a starting point for DNA replication. In most natural DNA replication, the ultimate primer for DNA synthesis is a short strand of RNA.
Biology	Biology is the branch of science dealing with the study of life. It is concerned with the characteristics, classification, and behaviors of organisms, how species come into existence, and the interactions they have with each other and with the environment.
Brown algae	One of a group of marine, multicellular, autotrophic protists, the most common type of seaweed. Brown algae include the kelps.
Sea anemone	Named after a terrestrial flower, the sea anemone is a group of water-dwelling, filter feeding animals of the order Actinaria. As a cnidarian, it is closely related to coral and

	jellyfish.
Symbiosis	Symbiosis is an interaction between two organisms living together in more or less intimate association or even the merging of two dissimilar organisms. The term host is usually used for the larger (macro) of the two members of a symbiosis.
Demography	Demography refers to the statistical study of population. It encompasses the study of the size, structure and distribution of populations, and how populations change over time due to births, deaths, migration and ageing.
Sexual dimorphism	Sexual dimorphism is the systematic difference in form between individuals of different sex in the same species.
Parasitoid	A parasitoid spends a significant portion of its life history attached to or within a single host organism which it ultimately kills and consumes. Thus they are similar to parasites except in the fate of the host.
Salt marsh	Salt marsh refers to a grassy area that extends along the shores of estuaries and sheltered coasts in temperate and subpolar regions.
Population structure	Population structure refers to the proportions of individuals in a population belonging to different age classes. Also, the distribution of the population in space.
Phylogeny	The science that explores the evolutionary relationships among organisms and seeks to reconstruct evolutionary history is phylogeny.
Sperm	Sperm refers to the male sex cell with three distinct parts at maturity: head, middle piece, and tail.
Polychaeta	The Polychaeta are a class of annelid worms, generally marine, with a pair of fleshy protrusions on each body segment called parapodia that bear many bristles, called setae, which are made of Chitin.
Pollution	Any environmental change that adversely affects the lives and health of living things is referred to as pollution.
Systematics	In biology, systematics is the study of the diversity of organism characteristics, and especially how they relate evolutionarily. The term "systematics" may or may not overlap with "taxonomy" which concerns itself with scientific classification of species and other taxa.
Lacuna	Small pit or hollow cavity, as in bone or cartilage, where a cell or cells are located is referred to as the lacuna.
Intermediate disturbance hypothesis	Intermediate disturbance hypothesis refers to the idea that species diversity is greatest in places where disturbance is moderate in both severity and frequency.
Intraspecific competition	Competition between individuals of the same species in a community for a limited resource is referred to as intraspecific competition.
Ecological succession	Ecological succession refers to the process of biological community change resulting from disturbance; transition in the species composition of a biological community, often following a flood, fire, or volcanic eruption.

Salinity	Salinity refers to the total amount of salts dissolved in seawater. It is generally expressed in parts per thousand.
Deep sea	The deep sea refers to those areas of oceans to which little or no light penetrates.
Habitat	Habitat refers to a place where an organism lives; an environmental situation in which an organism lives.
Tide pool	A depression that holds seawater at low tide is referred to as a tide pool.
Estuary	An estuary is a semi-enclosed coastal body of water which has a free connection with the open sea and within which sea water mixes with fresh water. The key feature of an estuary is that it is a mixing place for sea water and a stream or river to supply fresh water.
Tide	The periodic, rhythmic rise and fall of the sea surface caused by changes in gravitational forces external to the Earth is referred to as the tide.
Gene	Gene refers to a discrete unit of hereditary information consisting of a specific nucleotide sequence in DNA . Most of the genes of a eukaryote are located in its chromosomal DNA; a few are carried by the DNA of mitochondria and chloroplasts.
Receptor	A receptor is a protein on the cell membrane or within the cytoplasm or cell nucleus that binds to a specific molecule (a ligand), such as a neurotransmitter, hormone, or other substance, and initiates the cellular response to the ligand. Receptor, in immunology, the region of an antibody which shows recognition of an antigen.
Snail	The name snail applies to most members of the molluscan class Gastropoda that have coiled shells. Other gastropods, which lack a conspicuous shell, are commonly called slugs, and are scattered throughout groups that primarily include snails.
Protein	A protein is a complex, high-molecular-weight organic compound that consists of amino acids joined by peptide bonds. They are essential to the structure and function of all living cells and viruses. Many are enzymes or subunits of enzymes.
Molecule	A molecule is the smallest particle of a pure chemical substance that still retains its chemical composition and properties.
Hormone	A hormone is a chemical messenger from one cell to another. All multicellular organisms produce hormones. The best known hormones are those produced by endocrine glands of vertebrate animals, but hormones are produced by nearly every organ system and tissue type in a human or animal body. Hormone molecules are secreted directly into the bloodstream, they move by circulation or diffusion to their target cells, which may be nearby cells in the same tissue or cells of a distant organ of the body.
Enzyme	An enzyme is a protein that catalyzes, or speeds up, a chemical reaction. They are essential to sustain life because most chemical reactions in biological cells would occur too slowly, or would lead to different products, without them.
Metabolic rate	Energy expended by the body per unit time is called metabolic rate.
Reproduction	Biological reproduction is the biological process by which new individual organisms are produced. Reproduction is a fundamental feature of all known life; each individual organism exists as the result of reproduction by an antecedent.
Metabolism	Metabolism is the biochemical modification of chemical compounds in living organisms and cells. This includes the biosynthesis of complex organic molecules (anabolism) and their breakdown (catabolism).
Acclimation	Acclimation describes the process of an organism being forced to adjust to changes in their environment by artificial means, such as in a laboratory setting

Steady state	Steady state is a system in which a particular variable is not changing but energy must be continuously added to maintain this variable constant.
Adaptation	A biological adaptation is an anatomical structure, physiological process or behavioral trait of an organism that has evolved over a period of time by the process of natural selection such that it increases the expected long-term reproductive success of the organism.
Conformer	A characterization of an organism in regard to environmental variables. The organism is a conformer if it allows some conditions within its body to vary with certain external changes.
Salt	Salt is a term used for ionic compounds composed of positively charged cations and negatively charged anions, so that the product is neutral and without a net charge.
Species	Group of similarly constructed organisms capable of interbreeding and producing fertile offspring is a species.
Gamete	A gamete is a specialized germ cell that unites with another gamete during fertilization in organisms that reproduce sexually. They are haploid cells; that is, they contain one complete set of chromosomes. When they unite they form a zygote—a cell having two complete sets of chromosomes and therefore diploid.
Respiration	Respiration is the process by which an organism obtains energy by reacting oxygen with glucose to give water, carbon dioxide and ATP (energy). Respiration takes place on a cellular level in the mitochondria of the cells and provide the cells with energy.
Somatic	The term somatic refers to the body. It also refers to the part of the nervous system that controls voluntary movement and senzation and judges relative effort and weight, called proprioception.
Chemical reaction	Chemical reaction refers to a process leading to chemical changes in matter; involves the making and/or breaking of chemical bonds.
Cell	The cell is the structural and functional unit of all living organisms, and is sometimes called the "building block of life."
Oxygen	Oxygen is a chemical element in the periodic table. It has the symbol O and atomic number 8. Oxygen is the second most common element on Earth, composing around 46% of the mass of Earth's crust and 28% of the mass of Earth as a whole, and is the third most common element in the universe.
Mussel	A mussel is a bivalve mollusk that can be found in lakes, rivers, creeks, intertidal areas, and throughout the ocean. The saltwater mussels (family Mytilidae) and freshwater mussels (family Unionidae) are not thought to be closely related and are grouped in different subclasses, despite considerable similarities in appearance.
Crab	The term crab is often applied to several different groups of short (nose to tail) decapod crustaceans with thick exoskeletons, but only members of the Brachyura are true.
Mortality rate	Mortality rate is the number of deaths (from a disease or in general) per 1000 people and typically reported on an annual basis.
Population	Group of organisms of the same species occupying a certain area and sharing a common gene pool is referred to as population.
Distribution	Distribution in pharmacology is a branch of pharmacokinetics describing reversible transfer of drug from one location to another within the body.
Poikilotherm	An animal whose body temperature tends to vary with the surrounding environment is a poikilotherm. They are cold blooded.
Homeotherm	Homeotherm refers to an organism that maintains a stable internal body temperature regardless

	of external influence. This temperature is often higher than the immediate environment
Conception	Conception is fusion of gametes to form a new organism. In animals, the process involves a sperm fusing with an ovum, which eventually leads to the development of an embryo.
Mammal	Homeothermic vertebrate characterized especially by the presence of hair and mammary glands is a mammal.
Style	The style is a stalk connecting the stigma with the ovary below containing the transmitting tract, which facilitates the movement of the male gamete to the ovule.
Invertebrate	Invertebrate is a term coined by Jean-Baptiste Lamarck to describe any animal without a spinal column. It therefore includes all animals except vertebrates (fish, reptiles, amphibians, birds and mammals).
Conductance	The capacity of heat, electricity, or a substance to pass through a particular material is a conductance.
Skin	Skin is an organ of the integumentary system composed of a layer of tissues that protect underlying muscles and organs.
Circulatory system	The circulatory system or cardiovascular system is the organ system which circulates blood around the body of most animals.
Blood	Blood is a circulating tissue composed of fluid plasma and cells. The main function of blood is to supply nutrients (oxygen, glucose) and constitutional elements to tissues and to remove waste products.
Countercurrent heat exchanger	Parallel blood vessels that convey warm and cold blood in opposite directions, maximizing heat transfer to the cold blood is called countercurrent heat exchanger.
Venous blood	In the circulatory system, venous blood or peripheral blood is blood returning to the heart. With one exception (the pulmonary vein) this blood is deoxygenated and high in carbon dioxide, having released oxygen and absorbed CO_2 in the tissues.
Artery	Vessel that takes blood away from the heart to the tissues and organs of the body is called an artery.
Vein	Vein in animals, is a vessel that returns blood to the heart. In plants, a vascular bundle in a leaf, composed of xylem and phloem.
Anatomy	Anatomy is the branch of biology that deals with the structure and organization of living things. It can be divided into animal anatomy (zootomy) and plant anatomy (phytonomy).
Carbohydrate	Carbohydrate is a chemical compound that contains oxygen, hydrogen, and carbon atoms. They consist of monosaccharide sugars of varying chain lengths and that have the general chemical formula $C_n(H_2O)_n$ or are derivatives of such.
Mole	The atomic weight of a substance, expressed in grams. One mole is defined as the mass of 6.0222 3 1023 atoms.
Gradient	Gradient refers to a difference in concentration, pressure, or electrical charge between two regions.
Natural selection	Natural selection is the process by which biological individuals that are endowed with favorable or deleterious traits end up reproducing more or less than other individuals that do not possess such traits.
Geographic range	The geographic area in which a population lives is a geographic range.
Genus	In biology, a genus is a taxonomic grouping. That is, in the classification of living organisms, a genus is considered to be distinct from other such genera. A genus has one or

	more species: if it has more than one species these are likely to be morphologically more similar than species belonging to different genera.
Adenosine triphosphate	Adenosine triphosphate is the nucleotide known in biochemistry as the "molecular currency" of intracellular energy transfer; that is, able to store and transport chemical energy within cells.
Mitochondria	Mitochondria are organelles found in most eukaryotic cells, including those of plants, animals, fungi, and protists. Mitochondria are sometimes described as "cellular power plants", because their primary function is to convert organic materials into energy in the form of ATP.
Biota	Biota is the plant and animal life of a region or area or geological period. The biota of the Earth is known as the biosphere.
Coral	Group of cnidarians having a calcium carbonate skeleton that participate in the formation of reefs is coral.
Larva	A free-living, sexually immature form in some animal life cycles that may differ from the adult in morphology, nutrition, and habitat is called larva.
Cnidaria	Cnidaria is a phylum containing some 10,000 species of relatively simple animals found exclusively in aquatic environments (most species are marine). They get their name from cnidocytes, which are specialized cells that carry stinging organelles.
Phylum	Phylum is the second highest taxonomic classification for the kingdom Animalia (animals), between kingdom level and class level.
Gastropod	A member of the largest group of mollusks, including snails and slugs is a gastropod.
Veliger	Veliger are the free-swimming, planktonic larvae of certain marine and fresh-water univalve mollusks, or gastropods. Swimming is accomplished by one or more ciliated membranes.
Calcareous	Calcareous refers to a sediment, sedimentary rock, or soil type which is formed from or contains a high proportion of calcium carbonate in the form of calcite or aragonite. It is also used to refer to relatively alkaline soil.
Sea urchin	A sea urchin is a spiny sea creature of the class Echinoidea found in oceans all over the world. Their shell, which biologists call the test, is globular in shape, and covered with spines.
Zooplankton	Nonphotosynthetic protists that feed on other plankton and are abundant in marine and freshwater environments are zooplankton.
Micrograph	A micrograph is a photograph or similar image taken through a microscope or similar device to show a magnified image of an item.
Plankton	Plankton are drifting organisms that inhabit the water column of oceans, seas, and bodies of fresh water. Plankton abundance and distribution are strongly dependent on factors such as ambient nutrients concentrations, the physical state of the water column, and the abundance of other plankton.
Electron	The electron is a light fundamental subatomic particle that carries a negative electric charge. The electron is a spin-1/2 lepton, does not participate in strong interactions and has no substructure.
Horizon	Horizon refers to levels within a soil profile that differ structurally and chemically. Generally divided into A, B, C, E, and 0 horizons.
Phytoplankton	Phytoplankton refers to the autotrophic component of the plankton that drifts in the water column. Phytoplankton obtain energy through a process called photosynthesis, and so must live

in the well-lit surface layer (termed the euphotic zone) of an ocean, sea, or lake.

Temperate In virology, temperate refers to the life cycle a phage is able to perform. The phage usually integrates its genome into its host chromosome, becoming a prophage.

Diatom Diatom refers to a unicellular photosynthetic alga with a unique, glassy cell wall containing silica.

Boreal Boreal is usually applied to ecosystems localized to subarctic (Northern hemisphere) and subantarctic (Southern hemisphere) zones, although Austral is also used for the latter. A boreal forest, also known as the taiga, is the set of forest ecosystems that can survive in northern, specifically subarctic, regions.

Ciliate Ciliate is one of the most important groups of protists, common almost everywhere there is water. They are a type of protozoan that moves by means of cilia.

Jellyfish Jellyfish are marine invertebrates belonging to the Scyphozoa class, and in turn the phylum Cnidaria. The body of an adult jellyfish is composed of a bell-shaped, jellylike substance enclosing its internal structure, from which the creature's tentacles suspend.

Continental slope Continental slope the part of the continental margin that lies between the continental shelf and the bottom of the ocean. Sunlight does not penetrate this area, and mostly it is home to scavengers. It is characterized by a relatively steep slope of 3° to 6°.

Predator A predator is an animal or other organism that hunts and kills other organisms for food in an act called predation.

Ctenophore A ctenophore is a jellyfish-like animal commonly called a comb jelly. Despite their superficial similarities with jellyfish, they have a much more complex anatomy.

Crustacean Member of a group of marine arthropods that contains among others shrimps, crabs, crayfish, and lobsters is called crustacean.

Copepods A group of small crustaceans that are important members of marine and freshwater plankton communities is copepods.

Copepod The group of small crustaceans that is an important member of marine and freshwater plankton communities is copepod. They are very important food organisms for small fish, whales, and other crustaceans such as krill in the ocean and in fresh water.

Primary production Primary production is the production of organic compounds from inorganic materials principally through the process of photosynthesis.

Substrate A substrate is a molecule which is acted upon by an enzyme. Each enzyme recognizes only the specific substrate of the reaction it catalyzes. A surface in or on which an organism lives.

Heat shock protein Heat shock protein refers to protein synthesized in increased amounts in response to an elevated temperature or other stressful treatment, and which usually helps the cell to survive the stress. Prominent examples are hsp60 and hsp70.

Molecular weight The molecular mass of a substance, called molecular weight and abbreviated as MW, is the mass of one molecule of that substance, relative to the unified atomic mass unit u (equal to 1/12 the mass of one atom of carbon-12).

Ubiquitin Ubiquitin is a small protein that occurs in all eukaryotic cells. Its main function is to mark other proteins for destruction, known as proteolysis. Several ubiquitin molecules attach to the condemned protein, and it then moves to a proteasome, a barrel-shaped structure where the proteolysis occurs.

Intertidal zone The part of a beach that lies between average high tide and average low tide is called the intertidal zone.

Metabolic processes	The total of all chemical reactions within an organism are the metabolic processes.
Cell membrane	A component of every biological cell, the selectively permeable cell membrane is a thin and structured bilayer of phospholipid and protein molecules that envelopes the cell. It separates a cell's interior from its surroundings and controls what moves in and out.
Phospholipid	Phospholipid is a class of lipids formed from four components: fatty acids, a negatively-charged phosphate group, an alcohol and a backbone. Phospholipids with a glycerol backbone are known as glycerophospholipids or phosphoglycerides.
Vesicle	In cell biology, a vesicle is a relatively small and enclosed compartment, separated from the cytosol by at least one lipid bilayer.
Respiratory system	The respiratory system is the biological system of any organism that engages in gas exchange.In humans and other mammals, the respiratory system consists of the airways, the lungs, and the respiratory muscles that mediate the movement of air into and out of the body.
Foraminifera	The Foraminifera, or forams for short, are a large group of amoeboid protists with reticulating pseudopods, fine strands that branch and merge to form a dynamic net. They typically produce a shell, or test, which can have either one or multiple chambers, some becoming quite elaborate in structure.
Barnacle	A barnacle is a type of arthropod belonging to infraclass Cirripedia in the subphylum Crustacea and is hence distantly related to crabs and lobsters. They live attached to surfaces and are typically enclosed by heavy calcareous plates are referred to as barnacles.
Algae	The algae consist of several different groups of living organisms that capture light energy through photosynthesis, converting inorganic substances into simple sugars with the captured energy.
Bony fish	Osteichthyes are the bony fish, a group paraphyletic to the land vertebrates, which are sometimes included. Most belong to the Actinopterygii.
Crystal	Crystal is a solid in which the constituent atoms, molecules, or ions are packed in a regularly ordered, repeating pattern extending in all three spatial dimensions.
Glycoprotein	A macromolecule consisting of one or more polypeptides linked to short chains of sugars is called glycoprotein.
Tissue fluid	Tissue fluid is a solution which bathes and surrounds the cells of multicellular animals and is also called the intercellular fluid.
Bivalve	Bivalve refers to a member of a group of mollusks that includes clams, mussels, scallops, and oysters. They typically have two-part shells, with both parts being more or less symmetrical.
Sexual reproduction	The propagation of organisms involving the union of gametes from two parents is sexual reproduction.
Gonad	Gonad refers to a sex organ in an animal; an ovary or a testis. It is the organ that makes gametes.
Sperm	Sperm refers to the male sex cell with three distinct parts at maturity: head, middle piece, and tail.
Egg	An egg is the zygote, resulting from fertilization of the ovum. It nourishes and protects the embryo.
Potassium	Potassium is a chemical element in the periodic table. It has the symbol K (L. kalium) and atomic number 19. Potassium is a soft silvery-white metallic alkali metal that occurs naturally bound to other elements in seawater and many minerals.

Sodium	Sodium is the chemical element in the periodic table that has the symbol Na (Natrium in Latin) and atomic number 11. Sodium is a soft, waxy, silvery reactive metal belonging to the alkali metals that is abundant in natural compounds (especially halite). It is highly reactive.
Diffusion	Diffusion refers to the spontaneous movement of particles of any kind from where they are more concentrated to where they are less concentrated.
Osmosis	Osmosis is the diffusion of a solvent through a semipermeable membrane from a region of low solute concentration to a region of high solute concentration.
Solute	Substance that is dissolved in a solvent, forming a solution is referred to as a solute.
Hyperosmotic	The condition in which a solution has a higher osmotic concentration than that of a second solution is called hyperosmotic.
Hypoosmotic	Hypoosmotic refers to the condition in which a solution has a lower osmotic concentration than that of a second solution.
Bryozoan	Colonial animals that build stony skeletons of calcium carbonate and superficially resemble coral is referred to as bryozoan.
Osmotic pressure	Osmotic pressure is the pressure produced by a solution in a space that is enclosed by a differentially permeable membrane.
Unicellular	Unicellular organisms carry out all the functions of life. Unicellular species are those whose members consist of a single cell throughout their life cycle. This latter qualification is significant since most multicellular organisms consist of a single cell at the beginning of their life cycles.
Glycerol	Glycerol is a three-carbon substance that forms the backbone of fatty acids in fats. When the body uses stored fat as a source of energy, glycerol and fatty acids are released into the bloodstream. The glycerol component can be converted to glucose by the liver and provides energy for cellular metabolism.
Seaweed	Phycologists consider seaweed to refer any of a large number of marine benthic algae that are multicellular, macrothallic (large-bodied), and thus differentiated from most algae that tend to be microscopic in size.
Sucrose	A disaccharide composed of glucose and fructose is called sucrose.
Amino acid	An amino acid is any molecule that contains both amino and carboxylic acid functional groups. They are the basic structural building units of proteins. They form short polymer chains called peptides or polypeptides which in turn form structures called proteins.
Bacteria	The domain that contains procaryotic cells with primarily diacyl glycerol diesters in their membranes and with bacterial rRNA. Bacteria also is a general term for organisms that are composed of procaryotic cells and are not multicellular.
Hagfish	A hagfish is a marine chordate of the class Myxini, also known as Hyperotreti. Despite their name, there is some debate about whether they are strictly fish, since they belong to a much more primitive lineage than any other fish group.
Mollusk	The mollusk is the large and diverse phylum Mollusca, which includes a variety of familiar creatures well-known for their decorative shells or as seafood. These range from tiny snails, clams, and abalone to the octopus, cuttlefish and squid (which are considered the most intelligent invertebrates).
Urea	Urea is an organic compound of carbon, nitrogen, oxygen and hydrogen, CON_2H_4 or $(NH_2)_2CO$. Urea is essentially a waste product: it has no physiological function. It is dissolved in blood and excreted by the kidney.

Evolution	In biology, evolution is the process by which novel traits arise in populations and are passed on from generation to generation. Its action over large stretches of time explains the origin of new species and ultimately the vast diversity of the biological world.
Secretion	Secretion is the process of segregating, elaborating, and releasing chemicals from a cell, or a secreted chemical substance or amount of substance.
Urine	Concentrated filtrate produced by the kidneys and excreted via the bladder is called urine.
Gill	An extension of the body surface of an aquatic animal, specialized for gas exchange and/or suspension feeding is called a gill.
Ion	Ion refers to an atom or molecule that has gained or lost one or more electrons, thus acquiring an electrical charge.
Sargasso sea	Sargasso sea refers to the region of the Atlantic Ocean north of the West Indies that is characterized by floating masses of Sargasso weed, a brown alga.
Salt marsh	Salt marsh refers to a grassy area that extends along the shores of estuaries and sheltered coasts in temperate and subpolar regions.
Protozoa	Protozoa are single-celled eukaryotes (organisms with nuclei) that show some characteristics usually associated with animals, most notably mobility and heterotrophy. They are often grouped in the kingdom Protista together with the plant-like algae.
Skeleton	In biology, the skeleton or skeletal system is the biological system providing physical support in living organisms.
Tissue	Group of similar cells which perform a common function is called tissue.
Organ	Organ refers to a structure consisting of several tissues adapted as a group to perform specific functions.
Sea squirts	Tunicates with a sac-like, attached body called a tunic as adults are referred to as sea squirts.
Cephalopod	Cephalopod refers to a member of a group of mollusks that includes squids and octopuses, characterized by bilateral body symmetry.
Sponge	An invertebrates that consist of a complex aggregation of cells, including collar cells, and has a skeleton of fibers and/or spicules is a sponge. They are primitive, sessile, mostly marine, waterdwelling filter feeders that pump water through their matrix to filter out particulates of food matter.
Obligate anaerobe	An organism that cannot survive in the presence of oxygen is called an obligate anaerobe.
Anaerobic	An anaerobic organism is any organism that does not require oxygen for growth.
Anaerobe	An organism whose respiration does not require oxygen is an anaerobe.
Aerobe	An aerobic organism or aerobe is an organism that has an oxygen based metabolism. Aerobes, in a process known as cellular respiration, use oxygen to oxidize substrates (for example sugars and fats) in order to obtain energy.
Organelle	In cell biology, an organelle is one of several structures with specialized functions, suspended in the cytoplasm of a eukaryotic cell.
Flagella	Flagella are whip-like organelle that many unicellular organisms, and some multicellular ones, use to move about.
Nucleus	In cell biology, the nucleus is found in all eukaryotic cells that contains most of the cell's genetic material. The nucleus has two primary functions: to control chemical reactions

	within the cytoplasm and to store information needed for cellular division.
Cilia	Numerous short, hairlike structures projecting from the cell surface that enable locomotion are cilia.
Interstitial water	The water contained between sediment particles is interstitial water.
Sediment	Sediment is any particulate matter that can be transported by fluid flow and which eventually is deposited as a layer of solid particles on the bed or bottom of a body of water or other liquid.
Nutrition	Nutrition refers to collectively, the processes involved in taking in, assimilating, and utilizing nutrients.
Muscle contraction	A muscle contraction occurs when a muscle cell (called a muscle fiber) shortens. There are three general types: skeletal, heart, and smooth.
Vertebrate	Vertebrate is a subphylum of chordates, specifically, those with backbones or spinal columns. They started to evolve about 530 million years ago during the Cambrian explosion, which is part of the Cambrian period.
Glycogen	Glycogen refers to a complex, extensively branched polysaccharide of many glucose monomers; serves as an energy-storage molecule in liver and muscle cells.
Glucose	Glucose, a simple monosaccharide sugar, is one of the most important carbohydrates and is used as a source of energy in animals and plants. Glucose is one of the main products of photosynthesis and starts respiration.
Lactic acid	Lactic acid accumulates in skeletal muscles during extensive anaerobic exercise, causing temporary muscle pain. Lactic acid is quickly removed from muscles when they resume aerobic metabolism.
Glycolysis	Glycolysis refers to the multistep chemical breakdown of a molecule of glucose into two molecules of pyruvic acid; the first stage of cellular respiration in all organisms; occurs in the cytoplasmic fluid.
Muscle	Muscle is a contractile form of tissue. It is one of the four major tissue types, the other three being epithelium, connective tissue and nervous tissue. Muscle contraction is used to move parts of the body, as well as to move substances within the body.
Alanine	Alanine (Ala) also 2-aminopropanoic acid is a non-essential α-amino acid. It exists as two distinct enantiomers - L-alanine and D-alanine. L-alanine is one of the 20 amino acids most widely used in protein synthesis, second to leucine.
Acid	An acid is a water-soluble, sour-tasting chemical compound that when dissolved in water, gives a solution with a pH of less than 7.
Calcium carbonate	A mineral that is the major component of the shell, skeleton, and other parts of many organisms are called calcium carbonate.
Embryo	Embryo refers to a developing stage of a multicellular organism. In humans, the stage in the development of offspring from the first division of the zygote until body structures begin to appear, about the ninth week of gestation.
Crustacea	The animal phylum that includes mostly aquatic animals such as crabs, lobsters, crayfish, and shrimp is referred to as crustacea.
Respiratory surface	Respiratory surface refers to the part of an animal where gases are exchanged with the environment.
Lungs	Lungs are the essential organs of respiration in air-breathing vertebrates. Their principal

	function is to transport oxygen from the atmosphere into the bloodstream, and to excrete carbon dioxide from the bloodstream into the atmosphere.
Threshold	Electrical potential level at which an action potential or nerve impulse is produced is called threshold.
Ventilation	Ventilation refers to a mechanism that provides contact between an animal's respiratory surface and the air or water to which it is exposed. It is also called breathing.
Enhancer	An enhancer is a short region of DNA that can be bound with proteins to enhance transcription levels of genes in a gene-cluster.
Pigment	Pigment is any material resulting in color in plant or animal cells which is the result of selective absorption.
Hemoglobin	Hemoglobin is the iron-containing oxygen-transport metalloprotein in the red cells of the blood in mammals and other animals. Hemoglobin transports oxygen from the lungs to the rest of the body, such as to the muscles, where it releases the oxygen load.
Iron	Iron is essential to all organisms, except for a few bacteria. It is mostly stably incorporated in the inside of metalloproteins, because in exposed or in free form it causes production of free radicals that are generally toxic to cells.
Arthropod	Arthropod refers to member of a phylum of invertebrates that contains among other groups crustaceans and insects that have an exoskeleton and jointed appendages.
Brachiopod	A brachiopod is an animala that superficially resembles clams and other bivalve mollusks, but the two halves of the brachiopod shell are dorsal and ventral to the animal rather than lateral, as in clams.
Porphyrin	A porphyrin is a heterocyclic macrocycle made from 4 pyrrole subunits linked on opposite sides (á position) through 4 methine bridges (=CH-).
Echinoderm	Echinoderm is a phylum of marine animals found in the ocean at all depths. This phylum dates back to the lower Cambrian period and represents about 7000 living species and 13000 extinct ones. 6 classes made it to the Modern Era.
Flatworm	The flatworm is a phylum of relatively simple soft-bodied invertebrate animals. With about 25,000 known species they are the largest phylum of acoelomates. They are found in marine, freshwater, and even damp terrestrial environments.
Annelid	Annelid refers to member of a phylum of invertebrates that contains segmented worms, such as the earthworm and the clam worm.
Dissociation curve	A chart showing the relative amounts of oxygen bound to hemoglobin when the pigment is exposed to solutions varying in their partial pressure of dissolved oxygen is referred to as dissociation curve.
Dissolved oxygen	Oxygen gas molecules dissolved in water. Fish and other aquatic organisms are dependent on dissolved oxygen for respiration. Therefore concentration of dissolved oxygen is a measure of water quality.
Respiratory pigment	A respiratory pigment is an agent that will increase the oxygen-carrying capacity of the blood, such as the hemoglobin (red in color) in humans. The four most common invertebrate respiratory pigments are haemoglobin, haemocyanin, haemerythrin and chlorocruorin.
Carbon dioxide	Carbon dioxide is an atmospheric gas comprized of one carbon and two oxygen atoms. A very widely known chemical compound, it is frequently called by its formula CO_2. In its solid state, it is commonly known as dry ice.
Hemocyanin	Hemocyanin is a respiratory protein containing two copper atoms that reversibly bind a single

	oxygen molecule (O_2). Oxygenation causes a color change between the colorless Cu(I) deoxygenated form and the blue Cu(II) oxygenated form. Hemocyanins carry oxygen in the blood of most molluscs, and some arthropods such as the horseshoe crab.
Calcium	Calcium is the chemical element in the periodic table that has the symbol Ca and atomic number 20. Calcium is a soft grey alkaline earth metal that is used as a reducing agent in the extraction of thorium, zirconium and uranium. Calcium is also the fifth most abundant element in the Earth's crust.
Oxygen minimum layer	Oxygen minimum layer refers to a layer of water at a depth of approximately 500 m where the concentration of dissolved oxygen is very low.
Mantle cavity	A mantle skirt is a double fold of mantle that encloses a water space. This space is called the mantle cavity, and it is a central feature of mollusk biology, containing the mollusk's gills, anus, osphradium, nephridiopores, and gonopores.
Thermocline	The thermocline is a layer within a body of water where the temperature changes rapidly with depth. Because water is not perfectly transparent, almost all sunlight is absorbed in the surface layer, which heats up.
Migration	Migration occurs when living things move from one biome to another. In most cases organisms migrate to avoid local shortages of food, usually caused by winter. Animals may also migrate to a certain location to breed, as is the case with some fish.
Photosynthesis	Photosynthesis is a biochemical process in which plants, algae, and some bacteria harness the energy of light to produce food. Ultimately, nearly all living things depend on energy produced from photosynthesis for their nourishment, making it vital to life on Earth.
Surface layer	The upper layer of water that is mixed by wind, waves, and currents is the surface layer .
Biology	Biology is the branch of science dealing with the study of life. It is concerned with the characteristics, classification, and behaviors of organisms, how species come into existence, and the interactions they have with each other and with the environment.
Eye	An eye is an organ that detects light. Different kinds of light-sensitive organs are found in a variety of creatures.
Lens	The lens or crystalline lens is a transparent, biconvex structure in the eye that, along with the cornea, helps to refract light to focus on the retina. Its function is thus similar to a man-made optical lens.
Retina	The retina is a thin layer of cells at the back of the eyeball of vertebrates and some cephalopods; it is the part of the eye which converts light into nervous signals.
Cornea	The cornea is the transparent front part of the eye that covers the iris, pupil, and anterior chamber and provides most of an eye's optical power.
Compound eye	Compound eye refers to a type of eye, found in arthropods, that is composed of numerous independent subunits called ommatidia. Each ommatidium apparently contributes a piece of a mosaiclike image perceived by the animal.
Nervous system	The nervous system of an animal coordinates the activity of the muscles, monitors the organs, constructs and processes input from the senses, and initiates actions.
Mosaic	An organism containing tissues of different genotypes is referred to as mosaic.
Brain	The part of the central nervous system involved in regulating and controlling body activity and interpreting information from the senses transmitted through the nervous system is referred to as the brain.
Coral reef	A coral reef is a type of biotic reef that develops in tropical waters by the growth of

	coralline algae, hermatypic corals, and other marine organisms.
Spawning	A method of external fertilization in which male and female parents shed gametes into the water, and sperm must swim through the water to reach the eggs is referred to as spawning.
Bioluminescence	Bioluminescence is the production and emission of light by a living organism as the result of a chemical reaction during which chemical energy is converted to light energy.
Deep ocean	The deep ocean is the lowest layer in an ocean, existing below the thermocline. The deep ocean is not well mixed, consists of horizontal layers of equal density, and is often as cold as -1 to 4 degrees Celsius. Ninety percent of the total volume of Earth's oceans is found in the deep ocean.
Photophore	An organ that produces bioluminescence is called a photophore.
Tentacle	An elongate, extensible projection of the body of cnidarians and cephalopod mollusks that may be used for grasping, stinging, and immobilizing prey, and for locomotion is called tentacle.
Dinoflagellate	A unicellular photosynthetic alga with two flagella situated in perpendicular grooves in cellulose plates covering the cell is a dinoflagellate.
Sunlight	Sunlight in the broad sense is the total spectrum of electromagnetic radiation given off by the Sun.
Ventral	The surface or side of the body normally oriented upwards, away from the pull of gravity, is the dorsal side; the opposite side, typically the one closest to the ground when walking on all legs, swimming or flying, is the ventral side.
Diurnal	Diurnal is a term that refers to being active in the daytime.
Polar region	An area surrounding one of the two magnetic poles of a planet is a polar region.
Physiology	The study of the function of cells, tissues, and organs is referred to as physiology.
Aerobic metabolism	Aerobic metabolism is rather more efficient than anaerobic metabolism. They share the initial pathway of glycolysis but aerobic metabolism continues with the Krebs cycle and oxidative phosphorylation. The post glycolytic reactions take place in the mitochondria in eukaryotic cells, and at the cell membrane in prokaryotic cells.
Morphology	The scientific study of organic form, including both its development and function is morphology.
Matrix	In biology, matrix (plural: matrices) is the material between animal or plant cells, the material (or tissue) in which more specialized structures are embedded, and a specific part of the mitochondrion that is the site of oxidation of organic molecules.
Ecology	Ecology is the scientific study of the distribution and abundance of living organisms and how these properties are affected by interactions between the organisms and their environment.
Echinodermata	Echinodermata refers to the phylum of echinoderms, including sea stars, sea urchins, and sand dollars; characterized by a rough or spiny skin, a water vascular system, an endoskeleton, and radial symmetry in adults.
Bathypelagic	Bathypelagic is from 1000 m down to around 4000 m, by this depth the ocean is almost entirely dark. There are no living plants, and most animals survive by consuming the snow of detritus falling from the zones above, or by preying upon others. Giant squid live at this depth, and here they are hunted by deep-diving sperm whales.
Siphon	The tube-like extension through which water flows in and out of the mantle cavity in bivalves and cephalopods, and in tunicates is called siphon.
Biotic community	An association of plants, animals, and other organisms living together in a specific habitat

is a biotic community.

Sulfide One of the minerals that is abundant in the hot water that seeps through hydrothermal vents is sulfide.

Ubiquitination Ubiquitination refers to the process particular to eukaryotes whereby a protein is post-translationally modified by covalent attachment of a small protein.

Denaturation Denaturation is a structural change in biomolecules such as nucleic acids and proteins, such that they are no longer in their native state, and their shape which allows for optimal activity.

Reproduction | Biological reproduction is the biological process by which new individual organisms are produced. Reproduction is a fundamental feature of all known life; each individual organism exists as the result of reproduction by an antecedent.

Population | Group of organisms of the same species occupying a certain area and sharing a common gene pool is referred to as population.

Dispersal | Biological dispersal refers to those processes by which a species maintains or expands the distribution of a population. Dispersal implies movement—movement away from an existing population or away from the parent organisms.

Migration | Migration occurs when living things move from one biome to another. In most cases organisms migrate to avoid local shortages of food, usually caused by winter. Animals may also migrate to a certain location to breed, as is the case with some fish.

Habitat | Habitat refers to a place where an organism lives; an environmental situation in which an organism lives.

Species | Group of similarly constructed organisms capable of interbreeding and producing fertile offspring is a species.

Spawning | A method of external fertilization in which male and female parents shed gametes into the water, and sperm must swim through the water to reach the eggs is referred to as spawning.

Sexual reproduction | The propagation of organisms involving the union of gametes from two parents is sexual reproduction.

Zygote | Diploid cell formed by the union of sperm and egg is referred to as zygote.

Gamete | A gamete is a specialized germ cell that unites with another gamete during fertilization in organisms that reproduce sexually. They are haploid cells; that is, they contain one complete set of chromosomes. When they unite they form a zygote—a cell having two complete sets of chromosomes and therefore diploid.

Cell | The cell is the structural and functional unit of all living organisms, and is sometimes called the "building block of life."

Bacteria | The domain that contains procaryotic cells with primarily diacyl glycerol diesters in their membranes and with bacterial rRNA. Bacteria also is a general term for organisms that are composed of procaryotic cells and are not multicellular.

Mendelian inheritance | Mendelian inheritance is a set of primary tenets relating to the transmission of hereditary characteristics from parent organisms to their children, and underly much of genetics.

Chromosome | A chromosome is, minimally, a very long, continuous piece of DNA, which contains many genes, regulatory elements and other intervening nucleotide sequences.

Diploid | Diploid cells have two copies (homologs) of each chromosome (both sex- and non-sex determining chromosomes), usually one from the mother and one from the father. Most somatic cells (body cells) of complex organisms are diploid.

Genus | In biology, a genus is a taxonomic grouping. That is, in the classification of living organisms, a genus is considered to be distinct from other such genera. A genus has one or more species: if it has more than one species these are likely to be morphologically more similar than species belonging to different genera.

Crab | The term crab is often applied to several different groups of short (nose to tail) decapod crustaceans with thick exoskeletons, but only members of the Brachyura are true.

Genetic diversity | Genetic diversity is a characteristic of ecosystems and gene pools that describes an attribute which is commonly held to be advantageous for survival -- that there are many

	different versions of otherwise similar organisms.
Mutation	Mutation refers to a change in the nucleotide sequence of DNA; the ultimate source of genetic diversity.
Clone	A group of genetically identical cells or organisms derived by asexual reproduction from a single parent is called a clone.
Meiosis	In biology, meiosis is the process that transforms one diploid cell into four haploid cells in eukaryotes in order to redistribute the diploid's cell's genome. Meiosis forms the basis of sexual reproduction and can only occur in eukaryotes.
Gene	Gene refers to a discrete unit of hereditary information consisting of a specific nucleotide sequence in DNA . Most of the genes of a eukaryote are located in its chromosomal DNA; a few are carried by the DNA of mitochondria and chloroplasts.
Crossing over	An essential element of meiosis occurring during prophase when nonsister chromatids exchange portions of DNA strands is called crossing over.
Allele	An allele is any one of a number of viable DNA codings of the same gene (sometimes the term refers to a non-gene sequence) occupying a given locus (position) on a chromosome.
Natural selection	Natural selection is the process by which biological individuals that are endowed with favorable or deleterious traits end up reproducing more or less than other individuals that do not possess such traits.
Dominance	Dominance is the state of having high social status relative to other individuals, who react submissively to dominant individuals.
Phenotype	The phenotype of an individual organism is either its total physical appearance and constitution or a specific manifestation of a trait, such as size or eye color, that varies between individuals. It is determined to some extent by genotype.
Snail	The name snail applies to most members of the molluscan class Gastropoda that have coiled shells. Other gastropods, which lack a conspicuous shell, are commonly called slugs, and are scattered throughout groups that primarily include snails.
Parasite	A parasite is an organism that spends a significant portion of its life in or on the living tissue of a host organism and which causes harm to the host without immediately killing it. They also commonly show highly specialized adaptations allowing them to exploit host resources.
Sexual selection	Changes in males and females, often due to male competition and female selectivity leading to increased fitness are called sexual selection.
Protozoa	Protozoa are single-celled eukaryotes (organisms with nuclei) that show some characteristics usually associated with animals, most notably mobility and heterotrophy. They are often grouped in the kingdom Protista together with the plant-like algae.
Sperm	Sperm refers to the male sex cell with three distinct parts at maturity: head, middle piece, and tail.
Egg	An egg is the zygote, resulting from fertilization of the ovum. It nourishes and protects the embryo.
Fertilization	Fertilization is fusion of gametes to form a new organism. In animals, the process involves a sperm fusing with an ovum, which eventually leads to the development of an embryo.
Recognition protein	A protein or glycoprotein protruding from the outside surface of a plasma membrane that identifies a cell as belonging to a particular species, to a specific individual of that species, and in many cases to one specific organ within the individual is a recognition

	protein.
Sea urchin	A sea urchin is a spiny sea creature of the class Echinoidea found in oceans all over the world. Their shell, which biologists call the test, is globular in shape, and covered with spines.
Receptor	A receptor is a protein on the cell membrane or within the cytoplasm or cell nucleus that binds to a specific molecule (a ligand), such as a neurotransmitter, hormone, or other substance, and initiates the cellular response to the ligand. Receptor, in immunology, the region of an antibody which shows recognition of an antigen.
Protein	A protein is a complex, high-molecular-weight organic compound that consists of amino acids joined by peptide bonds. They are essential to the structure and function of all living cells and viruses. Many are enzymes or subunits of enzymes.
Speciation	Speciation refers to the evolutionary process by which new biological species arise. All forms of speciation have actually taken place over the course of evolution, though it still remains a subject of debate as to the relative importance of each mechanism in driving biodiversity.
Evolution	In biology, evolution is the process by which novel traits arise in populations and are passed on from generation to generation. Its action over large stretches of time explains the origin of new species and ultimately the vast diversity of the biological world.
Hermaphrodite	Hermaphrodite refers to an organism of a species whose members possess both male and female sexual organs during their lives. In many species, hermaphroditism is a normal part of the life-cycle. Generally, hermaphroditism occurs in the invertebrates, although it occurs in a fair number of fish, and to a lesser degree in other vertebrates.
Simultaneous hermaphrodite	Simultaneous hermaphrodite is when the organism has both male and female sexual organs at the same time as an adult. Usually, self-fertilization does not occur.
Self-fertilization	Self-fertilization occurs in hermaphroditic organisms where the two gametes fused in fertilization come from the same individual. This is common in plants and certain protozoans.
Hermaphroditism	The coexistence of both female and male sex organs in the same organism is called hermaphroditism.
Barnacle	A barnacle is a type of arthropod belonging to infraclass Cirripedia in the subphylum Crustacea and is hence distantly related to crabs and lobsters. They live attached to surfaces and are typically enclosed by heavy calcareous plates are referred to as barnacles.
Penis	The penis is the male reproductive organ and for mammals additionally serves as the external male organ of urination.
Sequential hermaphrodite	Sequential hermaphrodite is when the organism is born as one sex and later changes into the other sex.
Protogynous	Protogynous animals are sequential hermaphrodites, where that animal is first biologically female, having only female sexual organs, then changes to be biologically male.
Invertebrate	Invertebrate is a term coined by Jean-Baptiste Lamarck to describe any animal without a spinal column. It therefore includes all animals except vertebrates (fish, reptiles, amphibians, birds and mammals).
Coral reef	A coral reef is a type of biotic reef that develops in tropical waters by the growth of coralline algae, hermatypic corals, and other marine organisms.
Spermatocyte	Spermatocyte refers to the cell that gives rise to the sperm in animals. It divides by mitosis into two so-called primary spermatocytes.

Oocyte	An oocyte is a female gametocyte. Such that an oocyte is large and essentially stationary. The oocyte becomes functional when a lala (male gametocyte) attaches to it, thus allowing the meiosis of the secondary oocyte to occur.
Transformation	Transformation is the genetic alteration of a cell resulting from the introduction, uptake and expression of foreign genetic material (DNA or RNA).
Threshold	Electrical potential level at which an action potential or nerve impulse is produced is called threshold.
Trait	In biology, a trait or character is a genetically inherited feature of an organism: Eye color is a character; brown eyes and blue eyes is a trait.
Temperate	In virology, temperate refers to the life cycle a phage is able to perform. The phage usually integrates its genome into its host chromosome, becoming a prophage.
Starfish	Starfish are marine invertebrates belonging to phylum Echinodermata, class Asteroidea. They exhibit a superficially radial symmetry, typically with five or more "arms" which radiate from an indistinct disk. They do not have movable skeletons, but instead possess a hydraulic water vascular system.
Mussel	A mussel is a bivalve mollusk that can be found in lakes, rivers, creeks, intertidal areas, and throughout the ocean. The saltwater mussels (family Mytilidae) and freshwater mussels (family Unionidae) are not thought to be closely related and are grouped in different subclasses, despite considerable similarities in appearance.
Reproductive isolation	Reproductive isolation refers to the failure of organisms of one population to breed successfully with members of another; may be due to premating or postmating isolating mechanisms.
Specificity	A medical diagnostic test for a certain disease, specificity is the proportion of true negatives of all the negative samples tested.
Population density	The number of organisms of a species per unit area is called population density.
Bryozoan	Colonial animals that build stony skeletons of calcium carbonate and superficially resemble coral is referred to as bryozoan.
Biology	Biology is the branch of science dealing with the study of life. It is concerned with the characteristics, classification, and behaviors of organisms, how species come into existence, and the interactions they have with each other and with the environment.
Plasma membrane	Membrane surrounding the cytoplasm that consists of a phospholipid bilayer with embedded proteins is referred to as plasma membrane.
Geographic range	The geographic area in which a population lives is a geographic range.
Hybrid	Hybrid refers to the offspring of parents of two different species or of two different varieties of one species; the offspring of two parents that differ in one or more inherited traits; an individual that is heterozygous for one or more pair of genes.
Plankton	Plankton are drifting organisms that inhabit the water column of oceans, seas, and bodies of fresh water. Plankton abundance and distribution are strongly dependent on factors such as ambient nutrients concentrations, the physical state of the water column, and the abundance of other plankton.
Genetic marker	A genetic marker is a specific piece of DNA with a known position on the genome. It is a genetic technique to follow a certain disease or gene. It is known that pieces of DNA that lie near each other, tend to be inherited to the next organism together.

DNA sequence	A DNA sequence is a succession of letters representing the primary structure of a real or hypothetical DNA molecule or strand, The possible letters are A, C, G, and T, representing the four nucleotide subunits of a DNA strand (adenine, cytosine, guanine, thymine).
Amino acid	An amino acid is any molecule that contains both amino and carboxylic acid functional groups. They are the basic structural building units of proteins. They form short polymer chains called peptides or polypeptides which in turn form structures called proteins.
Molecule	A molecule is the smallest particle of a pure chemical substance that still retains its chemical composition and properties.
Concerted evolution	Maintenance of a homogeneous nucleotide sequence among the members of a gene family, which sequence evolves over time is called concerted evolution.
Polymorphism	The presence in a population of more than one allele of a gene at a frequency greater than that of newly arising mutations is referred to as polymorphism.
Mantle cavity	A mantle skirt is a double fold of mantle that encloses a water space. This space is called the mantle cavity, and it is a central feature of mollusk biology, containing the mollusk's gills, anus, osphradium, nephridiopores, and gonopores.
Gemma	A gemma is a single cell, or a mass of cells, or a modified bud of tissue, that detaches from the parent and develops into a new individual. It is a means of asexual propagation in plants.
Asexual reproduction	Asexual reproduction refers to reproduction that does not involve the fusion of haploid sex cells. The parent body may divide and new parts regenerate, or a new, smaller individual may form as an attachment to the parent, to drop off when complete.
Genotype	The genotype is the specific genetic makeup (the specific genome) of an individual, usually in the form of DNA. It codes for the phenotype of that individual.
Style	The style is a stalk connecting the stigma with the ovary below containing the transmitting tract, which facilitates the movement of the male gamete to the ovule.
Phytoplankton	Phytoplankton refers to the autotrophic component of the plankton that drifts in the water column. Phytoplankton obtain energy through a process called photosynthesis, and so must live in the well-lit surface layer (termed the euphotic zone) of an ocean, sea, or lake.
Fission	A means of asexual reproduction whereby a parent separates into two or more genetically identical individuals of about equal size is referred to as fission.
Diatom	Diatom refers to a unicellular photosynthetic alga with a unique, glassy cell wall containing silica.
Unicellular	Unicellular organisms carry out all the functions of life. Unicellular species are those whose members consist of a single cell throughout their life cycle. This latter qualification is significant since most multicellular organisms consist of a single cell at the beginning of their life cycles.
Skeleton	In biology, the skeleton or skeletal system is the biological system providing physical support in living organisms.
Silica	Silica is a mineral similar to glass that is the major component of the cell wall, shell, or skeleton of many marine organisms.
Cell division	Cell division is the process by which a cell (called the parent cell) divides into two cells (called daughter cells). Cell division is usually a small segment of a larger cell cycle. In meiosis, however, a cell is permanently transformed and cannot divide again.
Daughter cell	A cell formed by cell division of a parent cell is a daughter cell.

Fragmentation | Fragmentation is a form of asexual reproduction where an organism is split into fragments. The splitting may or may not be intentional. Each of these fragments develops into a mature, fully grown new organism, and if the organism is split any further the process is repeated.

Multicellular | Multicellular organisms are those organisms consisting of more than one cell, and having differentiated cells that perform specialized functions. Most life that can be seen with the naked eye is multicellular, as are all animals (i.e. members of the kingdom Animalia) and plants (i.e. members of the kingdom Plantae).

Annelid | Annelid refers to member of a phylum of invertebrates that contains segmented worms, such as the earthworm and the clam worm.

Seaweed | Phycologists consider seaweed to refer any of a large number of marine benthic algae that are multicellular, macrothallic (large-bodied), and thus differentiated from most algae that tend to be microscopic in size.

Coral | Group of cnidarians having a calcium carbonate skeleton that participate in the formation of reefs is coral.

Colony | Colony refers to a cluster or assemblage of microorganisms growing on the surface of an agar culture medium. A colony also refers to several individual organisms of the same species living closely together, usually for mutual benefit, such as stronger defences, the ability to attack bigger prey, etc.

Angiosperm | Flowering plant that produces seeds within an ovary that develops into a fruit is referred to as an angiosperm.

Algae | The algae consist of several different groups of living organisms that capture light energy through photosynthesis, converting inorganic substances into simple sugars with the captured energy.

Rhizome | A rhizome is a usually-underground, horizontal stem of a plant that often sends out roots and shoots from its nodes.

Sponge | An invertebrates that consist of a complex aggregation of cells, including collar cells, and has a skeleton of fibers and/or spicules is a sponge. They are primitive, sessile, mostly marine, waterdwelling filter feeders that pump water through their matrix to filter out particulates of food matter.

Sediment | Sediment is any particulate matter that can be transported by fluid flow and which eventually is deposited as a layer of solid particles on the bed or bottom of a body of water or other liquid.

Larva | A free-living, sexually immature form in some animal life cycles that may differ from the adult in morphology, nutrition, and habitat is called larva.

Tissue | Group of similar cells which perform a common function is called tissue.

Enzyme | An enzyme is a protein that catalyzes, or speeds up, a chemical reaction. They are essential to sustain life because most chemical reactions in biological cells would occur too slowly, or would lead to different products, without them.

Tentacle | An elongate, extensible projection of the body of cnidarians and cephalopod mollusks that may be used for grasping, stinging, and immobilizing prey, and for locomotion is called tentacle.

Adaptation | A biological adaptation is an anatomical structure, physiological process or behavioral trait of an organism that has evolved over a period of time by the process of natural selection such that it increases the expected long-term reproductive success of the organism.

Inclusive fitness | Inclusive fitness encompasses conventional Darwinian fitness with the addition of behaviors that contribute to an organism's individual fitness through altruism. An organism's ultimate

	goal is to leave the maximum number of viable offspring possible, thereby keeping their genes present within a population.
Kin selection	Kin Selection is the phrase used to refer to changes in gene frequency driven by natural selection that can only be understood by looking at how biological relatives influence the fitness of each other.
Eusociality	Eusociality is the phenomenon of reproductive specialization found in some species of animals. There are two causes of eusociality: kin-selected altruism and high inbreeding.
Insect	An arthropod that usually has three body segments , three pairs of legs, and one or two pairs of wings is called an insect. They are the largest and (on land) most widely-distributed taxon within the phylum Arthropoda. They comprise the most diverse group of animals on the earth, with around 925,000 species described
Mole	The atomic weight of a substance, expressed in grams. One mole is defined as the mass of 6.0222 3 1023 atoms.
Crustacean	Member of a group of marine arthropods that contains among others shrimps, crabs, crayfish, and lobsters is called crustacean.
Abdomen	The abdomen is a part of the body. In humans, and in many other vertebrates, it is the region between the thorax and the pelvis. In fully developed insects, the abdomen is the third (or posterior) segment, after the head and thorax.
Evolutionary tree	The evolutionary tree of living things is currently based on ideas from cladistics; where more than two groups are shown in a single branch, there is disagreement about how they diverged. Hypothetically taxonomy would follow the tree whenever possible, but in many places it does not at present.
Reproductive effort	The proportion of energy or materials that an organism allocates to reproduction rather than to growth and maintenance is called reproductive effort.
Demography	Demography refers to the statistical study of population. It encompasses the study of the size, structure and distribution of populations, and how populations change over time due to births, deaths, migration and ageing.
Morphology	The scientific study of organic form, including both its development and function is morphology.
Iteroparity	A life history in which adults produce large numbers of offspring over many years is an iteroparity.
Semelparity	Semelparity refers to a life history in which adults have but a single reproductive opportunity to produce large numbers of offspring, such as the life history of the Pacific salmon.
Somatic	The term somatic refers to the body. It also refers to the part of the nervous system that controls voluntary movement and senzation and judges relative effort and weight, called proprioception.
Repeated reproduction	Repeated reproduction refers to a life history in which adults produce large numbers of offspring over many years.
Distribution	Distribution in pharmacology is a branch of pharmacokinetics describing reversible transfer of drug from one location to another within the body.
Mammal	Homeothermic vertebrate characterized especially by the presence of hair and mammary glands is a mammal.
Anadromous	Anadromous refers to marine fishes that migrate to fresh water to breed.

Lamprey	A lamprey is a jawless fish with a toothed, funnel-like sucking mouth, with which most species bore into the flesh of other fishes to suck their blood.
Catadromous	Catadromous fish live in fresh water and breed in sea.
Zooplankton	Nonphotosynthetic protists that feed on other plankton and are abundant in marine and freshwater environments are zooplankton.
Sargasso sea	Sargasso sea refers to the region of the Atlantic Ocean north of the West Indies that is characterized by floating masses of Sargasso weed, a brown alga.
Continental shelf	The shallow part of the seafloor immediately adjacent to a continent is called the continental shelf.
Vertebrae	Vertebrae are the individual bones that make up the vertebral column (aka spine) - a flexuous and flexible column.
Imprinting	Learning that is limited to a specific critical period in an animal's life and that is generally irreversible is called imprinting.
Fish migration	Fish migration occurs on a regular basis, on time scales ranging from daily to annual, and with distances ranging from a few meters to thousands of kilometers. The purpose usually relates to either feeding or breeding; in some cases the reason is still unknown.
Salt	Salt is a term used for ionic compounds composed of positively charged cations and negatively charged anions, so that the product is neutral and without a net charge.
Growth rate	A measure of the change in population size per individual per unit of time is called growth rate.
Relative abundance	Differences in the abundance of species within a community is relative abundance.
Water column	The vertical column of seawater that extends from the surface to the bottom is called water column.
Predator	A predator is an animal or other organism that hunts and kills other organisms for food in an act called predation.
Competent	Description of a state of cells able to take up DNA from the medium is called competent.
Zooxanthellae	Zooxanthellae refers to dinoflagellates and other algae that live as symbionts within coral.
Invasive species	The term invasive species refers to a subset of those species defined as introduced species or non-indigenous species. Invasive species can alter ecological relationships among native species and can affect ecosystem function, economic value of ecosystems, and human health.
Gastropod	A member of the largest group of mollusks, including snails and slugs is a gastropod.
Veliger	Veliger are the free-swimming, planktonic larvae of certain marine and fresh-water univalve mollusks, or gastropods. Swimming is accomplished by one or more ciliated membranes.
Prediction	Step of the scientific method which follows the formulation of a hypothesis and assists in creating the experimental design is referred to as prediction.
Barrier reef	A type of coral reef that develops at some distance from the coast creating a lagoon between itself and the coast is called a barrier reef.
Gyre	Gyre refers to a roughly circular pattern of ocean currents, formed because continents interrupt the currents' flow; rotates clockwise in the Northern Hemisphere and counterclockwise in the Southern Hemisphere.
Viviparous	Viviparous refers to reproduction in which eggs develop within the mother's body and young

	are born free-living.
Embryo	Embryo refers to a developing stage of a multicellular organism. In humans, the stage in the development of offspring from the first division of the zygote until body structures begin to appear, about the ninth week of gestation.
Oviduct	In oviparous animals (those that lay eggs), the passage from the ovaries to the outside of the body is known as the oviduct. The eggs travel along the oviduct.
Crinoids	Crinoids refers to the echinoderms with a small, cup-shaped body and feathery arms: the sea lilies and feather stars .
Gonad	Gonad refers to a sex organ in an animal; an ovary or a testis. It is the organ that makes gametes.
Capsule	A sticky layer that surrounds the bacterial cell wall, protects the cell surface, and sometimes helps glue the cell to surfaces is called the capsule. In botany, a capsule is a type of dry fruit as in the poppy, iris, foxglove, etc. as well as another term for the sporangium of mosses and hornworts.
Meiofauna	Meiofauna are small benthic invertebrates that live in both marine and fresh water environments. The term Meiofauna loosely defines a group of organisms by their size, larger than Microfauna but smaller than Macrofauna, rather than a taxonomic grouping.
Yolk	Dense nutrient material that is present in the egg of a bird or reptile is referred to as yolk.
Mollusk	The mollusk is the large and diverse phylum Mollusca, which includes a variety of familiar creatures well-known for their decorative shells or as seafood. These range from tiny snails, clams, and abalone to the octopus, cuttlefish and squid (which are considered the most intelligent invertebrates).
Metamorphosis	Metamorphosis is a process in biology by which an individual physically develops after birth or hatching, and involves significant change in form as well as growth and differentiation.
Tadpole	A tadpole is a larval frog, toad, salamander, newt, or caecilian. In this stage it breathes by means of external or internal gills, is at first lacking legs, and has a finlike tail with which it swims as most fish do, by lateral undulation.
Substratum	Substratum is the underlying layer, or base to which a lichen is fixed.
Cilia	Numerous short, hairlike structures projecting from the cell surface that enable locomotion are cilia.
Downstream	Downstream refers to a relative position in DNA or RNA. Relative to the position on the strand, downstream is the region towards the 3' end of the strand.
Cilium	A cilium is an organelle projecting from a eukaryotic cell. They are extensions of the plasma membrane containing doublets of parallel microtubules.
Upstream	Upstream refers to movement opposite the direction RNA follows when moving along a gene.
Sea cucumber	The sea cucumber is an echinoderm of the class Holothuroidea, with an elongated body and leathery skin. Mostly found on the sea floor.
Pigment	Pigment is any material resulting in color in plant or animal cells which is the result of selective absorption.
Predation	Interaction in which one organism uses another, called the prey, as a food source is referred to as predation.
Tide	The periodic, rhythmic rise and fall of the sea surface caused by changes in gravitational

	forces external to the Earth is referred to as the tide.
Surface layer	The upper layer of water that is mixed by wind, waves, and currents is the surface layer .
Surface water	Includes all bodies of water, lakes, rivers, ponds, and so on that are on the surface of the earth, in contrast to groundwater, which lies below the surface is surface water.
Brittle star	A brittle star is an echinoderm with five flexible arms that radiate from a conspicuous central disk and tube feet that are used in feeding.
Brachiopod	A brachiopod is an animala that superficially resembles clams and other bivalve mollusks, but the two halves of the brachiopod shell are dorsal and ventral to the animal rather than lateral, as in clams.
Mortality rate	Mortality rate is the number of deaths (from a disease or in general) per 1000 people and typically reported on an annual basis.
Intertidal zone	The part of a beach that lies between average high tide and average low tide is called the intertidal zone.
Phototactic	Phototactic describes the movement of a cell (eg a gamete) or a motile organism towards or away from a light source.
Byssal thread	Byssal thread denotes strong threads secreted by mussels to attach to rocks and large, generally heavy objects in the intertidal zone. They range to 6 centimeters in length.
Organ	Organ refers to a structure consisting of several tissues adapted as a group to perform specific functions.
Filament	The stamen is the male organ of a flower. Each stamen generally has a stalk called the filament, and, on top of the filament, an anther. The filament is a long chain of proteins, such as those found in hair, muscle, or in flagella.
Deep sea	The deep sea refers to those areas of oceans to which little or no light penetrates.
Bivalve	Bivalve refers to a member of a group of mollusks that includes clams, mussels, scallops, and oysters. They typically have two-part shells, with both parts being more or less symmetrical.
Isotope	An isotope is a form of an element whose nuclei have the same atomic number - the number of protons in the nucleus - but different mass numbers because they contain different numbers of neutrons.
Oxygen	Oxygen is a chemical element in the periodic table. It has the symbol O and atomic number 8. Oxygen is the second most common element on Earth, composing around 46% of the mass of Earth's crust and 28% of the mass of Earth as a whole, and is the third most common element in the universe.
Peptide	Peptide is the family of molecules formed from the linking, in a defined order, of various amino acids. The link between one amino acid residue and the next is an amide bond, and is sometimes referred to as a peptide bond.
Boundary layer	Boundary layer refers to a layer of still or slow-moving water or air close to the surface of an object.
Red algae	One of a group of marine, mostly multicellular, autotrophic protists, which includes the reef-building coralline algae is referred to as red algae.
Aromatic	In chemistry, an aromatic molecule is one in which a conjugated ring of unsaturated bonds, lone pairs, or empty orbitals exhibit a stabilization stronger than would be expected by the stabilization of conjugation alone.
Net movement	Net movement refers to movement in one direction minus the movement in the other.

Estuary	An estuary is a semi-enclosed coastal body of water which has a free connection with the open sea and within which sea water mixes with fresh water. The key feature of an estuary is that it is a mixing place for sea water and a stream or river to supply fresh water.
Ctenophore	A ctenophore is a jellyfish-like animal commonly called a comb jelly. Despite their superficial similarities with jellyfish, they have a much more complex anatomy.
Crustacea	The animal phylum that includes mostly aquatic animals such as crabs, lobsters, crayfish, and shrimp is referred to as crustacea.
Spine	A spine is a rigid, pointed surface protuberance or needle-like structure on an animal, shell, or plant, presumably serving as a defense against attack by predators.
Primary production	Primary production is the production of organic compounds from inorganic materials principally through the process of photosynthesis.
Echinoderm	Echinoderm is a phylum of marine animals found in the ocean at all depths. This phylum dates back to the lower Cambrian period and represents about 7000 living species and 13000 extinct ones. 6 classes made it to the Modern Era.
Habitat selection	Choice of or preference for certain habitats is a habitat selection.
Residue	A residue refers to a portion of a larger molecule, a specific monomer of a polysaccharide, protein or nucleic acid.
Sea squirts	Tunicates with a sac-like, attached body called a tunic as adults are referred to as sea squirts.
Biogeography	Biogeography is the science which deals with patterns of species distribution and the processes that result in such patterns.
Upwelling	Upward movement of deep, nutrient-rich water along coasts is referred to as upwelling.
Conception	Conception is fusion of gametes to form a new organism. In animals, the process involves a sperm fusing with an ovum, which eventually leads to the development of an embryo.
Extinction	In biology and ecology, extinction is the ceasing of existence of a species or group of taxa. The moment of extinction is generally considered to be the death of the last individual of that species.The death of all members of a species is extinction.
Catastrophe	An unpredictable event that has a strong effect on a population is referred to as a catastrophe.
Perennial	Flowering plant that lives more than one growing season because the underground parts regrow each season is referred to as perennial.
Colonization	Colonization is the act where life forms move into a distant area where their kind is sparse or not yet existing at all and set up new settlements in the area. Colonization applies to all life forms in a sense though it is most often used in reference to insects and humans.
Diffusion	Diffusion refers to the spontaneous movement of particles of any kind from where they are more concentrated to where they are less concentrated.
Sequential hermaphroditism	Sequential hermaphroditism refers to a reproductive pattern in which an individual reverses its sex during its lifetime.
Mating system	The pattern of matings between individuals in a population, including number of simultaneous mates, permanence of pair bond, and degree of inbreeding is referred to as the mating system.
Behavioral ecology	Behavioral ecology is the study of the ecological and evolutionary basis for animal behavior, and the roles of behavior in enabling an animal to adapt to its environment.

Sociobiology | Application of evolutionary principles to the study of social behavior of animals, including humans is sociobiology.

Radiation | The emission of electromagnetic waves by all objects warmer than absolute zero is referred to as radiation.

Decapoda | The decapoda are an order of crustaceans within the class Malacostraca, including many familiar groups, such as crabs, lobsters, prawns and shrimps, but also some families that are less well known.

Territory | In ethology, sociobiology and behavioral ecology, the term territory refers to any geographical area that an animal of a particular species consistently defends against conspecifics (and, occasionally, animals of other species).

Sea anemone | Named after a terrestrial flower, the sea anemone is a group of water-dwelling, filter feeding animals of the order Actinaria. As a cnidarian, it is closely related to coral and jellyfish.

Echinodermata | Echinodermata refers to the phylum of echinoderms, including sea stars, sea urchins, and sand dollars; characterized by a rough or spiny skin, a water vascular system, an endoskeleton, and radial symmetry in adults.

Sexual dimorphism | Sexual dimorphism is the systematic difference in form between individuals of different sex in the same species.

Parasitism | A symbiotic relationship in which the parasite, a type of predator, lives within or on the surface of a host, from which it derives its food is a parasitism.

Rearrangement | A change in the usual order and arrangement of genetic material either within the chromosome complement or within a gene locus is rearrangement. Where the nature of the rearrangement has been determined, the type may be searched for directly under the following designations: reciprocal translocation, Robertsonian translocation, insertion, transposition, inversion, deletion, and duplication.

Fusion | Fusion refers to the combination of two atoms into a single atom as a result of a collision, usually accompanied by the release of energy.

Mitochondrial DNA | Mitochondrial DNA is DNA not located in the nucleus of the cell but in the mitochondria, parts of the cell that generate fuel in the form of adenosine triphosphate, which drives the varied machinery of the cell. Unlike most of the cell, the function of which is defined by the nuclear DNA, the mitochondria have their own DNA and are assumed to have evolved separately.

Sessile | In botany, sessile means without a stalk, as in flowers or leaves that grow directly from the stem. In zoology, sessile organisms are those which are not able to move about. They are usually permanently attached to a solid substrate of some kind, such as a rock, or the hull of a ship in the case of barnacles.

Polychaeta | The Polychaeta are a class of annelid worms, generally marine, with a pair of fleshy protrusions on each body segment called parapodia that bear many bristles, called setae, which are made of Chitin.

Kelp | Kelp are large seaweeds, belonging to the brown algae and classified in the order Laminariales. Despite their appearance they are not grouped with the normal aquatic or land plants (kingdom Plantae), but instead are included in either kingdom Protista or Chromista.

Infauna | Infauna are those organisms that live within particulate media such as sediments or soil. They are most common in the subtidal and deeper zones.

Ecology | Ecology is the scientific study of the distribution and abundance of living organisms and how

	these properties are affected by interactions between the organisms and their environment.
Mudflat	Mudflat is a coastal wetland that forms when mud is deposited by the tides or rivers, they are found in sheltered areas such as bays, bayous, lagoons, and estuaries.
Hydrothermal vent	A hydrothermal vent is a fissure in a planet's surface from which geothermally heated water issues. Hydrothermal vents are commonly found in places that are also volcanically active, where hot magma is relatively near the planet's surface.
Systematics	In biology, systematics is the study of the diversity of organism characteristics, and especially how they relate evolutionarily. The term "systematics" may or may not overlap with "taxonomy" which concerns itself with scientific classification of species and other taxa.
Tidal bore	Tidal bore refers to a steep wave generated as high tides move up some estuaries and rivers. tidal current. A current that is generated by tides.
Helix	A helix is a twisted shape like a spring, screw or a spiral staircase. They are important in biology, as DNA and many proteins have spiral substructures, known a alpha helix.
Genetics	Genetics is the science of genes, heredity, and the variation of organisms.

Oxygen	Oxygen is a chemical element in the periodic table. It has the symbol O and atomic number 8. Oxygen is the second most common element on Earth, composing around 46% of the mass of Earth's crust and 28% of the mass of Earth as a whole, and is the third most common element in the universe.
Suspension feeder	An animal that extracts food particles suspended in the surrounding water for food is referred to as a suspension feeder.
Species	Group of similarly constructed organisms capable of interbreeding and producing fertile offspring is a species.
Tissue	Group of similar cells which perform a common function is called tissue.
Salt	Salt is a term used for ionic compounds composed of positively charged cations and negatively charged anions, so that the product is neutral and without a net charge.
Threshold	Electrical potential level at which an action potential or nerve impulse is produced is called threshold.
Cilia	Numerous short, hairlike structures projecting from the cell surface that enable locomotion are cilia.
Molecule	A molecule is the smallest particle of a pure chemical substance that still retains its chemical composition and properties.
Boundary layer	Boundary layer refers to a layer of still or slow-moving water or air close to the surface of an object.
Copepods	A group of small crustaceans that are important members of marine and freshwater plankton communities is copepods.
Copepod	The group of small crustaceans that is an important member of marine and freshwater plankton communities is copepod. They are very important food organisms for small fish, whales, and other crustaceans such as krill in the ocean and in fresh water.
Bivalve	Bivalve refers to a member of a group of mollusks that includes clams, mussels, scallops, and oysters. They typically have two-part shells, with both parts being more or less symmetrical.
Fiber	Fiber is a class of materials that are continuous filaments or are in discrete elongated pieces, similar to lengths of thread. They are of great importance in the biology of both plants and animals, for holding tissues together.
Gill	An extension of the body surface of an aquatic animal, specialized for gas exchange and/or suspension feeding is called a gill.
Principle of continuity	The principle of continuity states that because life probably evolved from nonlife by a continuous, gradual process, all postulated stages in the evolution of life should be derivable from preexisting states.
Circulatory system	The circulatory system or cardiovascular system is the organ system which circulates blood around the body of most animals.
Choanocyte	Choanocyte refers to a flagellated feeding cell found in sponges. Also called a collar cell, it has a collarlike ring that traps food particles around the base of its flagellum.
Sponge	An invertebrates that consist of a complex aggregation of cells, including collar cells, and has a skeleton of fibers and/or spicules is a sponge. They are primitive, sessile, mostly marine, waterdwelling filter feeders that pump water through their matrix to filter out particulates of food matter.
Cell	The cell is the structural and functional unit of all living organisms, and is sometimes called the "building block of life."

Mollusk	The mollusk is the large and diverse phylum Mollusca, which includes a variety of familiar creatures well-known for their decorative shells or as seafood. These range from tiny snails, clams, and abalone to the octopus, cuttlefish and squid (which are considered the most intelligent invertebrates).
Siphon	The tube-like extension through which water flows in and out of the mantle cavity in bivalves and cephalopods, and in tunicates is called siphon.
Sediment	Sediment is any particulate matter that can be transported by fluid flow and which eventually is deposited as a layer of solid particles on the bed or bottom of a body of water or other liquid.
Skin	Skin is an organ of the integumentary system composed of a layer of tissues that protect underlying muscles and organs.
Downstream	Downstream refers to a relative position in DNA or RNA. Relative to the position on the strand, downstream is the region towards the 3' end of the strand.
Upstream	Upstream refers to movement opposite the direction RNA follows when moving along a gene.
Coral	Group of cnidarians having a calcium carbonate skeleton that participate in the formation of reefs is coral.
Sea anemone	Named after a terrestrial flower, the sea anemone is a group of water-dwelling, filter feeding animals of the order Actinaria. As a cnidarian, it is closely related to coral and jellyfish.
Seaweed	Phycologists consider seaweed to refer any of a large number of marine benthic algae that are multicellular, macrothallic (large-bodied), and thus differentiated from most algae that tend to be microscopic in size.
Tentacle	An elongate, extensible projection of the body of cnidarians and cephalopod mollusks that may be used for grasping, stinging, and immobilizing prey, and for locomotion is called tentacle.
Frond	A frond is the leaf structure of a fern. The term is colloquially applied to the leaves of palms, cycads, and other plants with pinnately compound leaves. A significant difference is that, unlike the leaves of the latter, fern fronds bear the reproductive structures (spore-bearing structures) of the sporophyte plant.
Invertebrate	Invertebrate is a term coined by Jean-Baptiste Lamarck to describe any animal without a spinal column. It therefore includes all animals except vertebrates (fish, reptiles, amphibians, birds and mammals).
Limpet	Limpet is a marine mollusk in the family Acmaeidae with flattened, cone-shaped shells. They live throughout the intertidal zone, from the high zone (upper littoral) to the shallow subtidal on the rocky coasts of most oceans. They can be commonly found attached to rocks, looking like little disks or bumps on the rock surface.
Snail	The name snail applies to most members of the molluscan class Gastropoda that have coiled shells. Other gastropods, which lack a conspicuous shell, are commonly called slugs, and are scattered throughout groups that primarily include snails.
Copper	Copper is a chemical element in the periodic table that has the symbol Cu (L.: Cuprum) and atomic number 29. It is a ductile metal with excellent electrical conductivity, and finds extensive use as a building material, as an electrical conductor, and as a component of various alloys.
Kinetic energy	Kinetic energy refers to energy that is actually doing work; the energy of a mass of matter that is moving. Moving matter performs work by transferring its motion to other matter, such as leg muscles pushing bicycle pedals.

Gradient	Gradient refers to a difference in concentration, pressure, or electrical charge between two regions.
Biology	Biology is the branch of science dealing with the study of life. It is concerned with the characteristics, classification, and behaviors of organisms, how species come into existence, and the interactions they have with each other and with the environment.
Substratum	Substratum is the underlying layer, or base to which a lichen is fixed.
Gastropod	A member of the largest group of mollusks, including snails and slugs is a gastropod.
Salinity	Salinity refers to the total amount of salts dissolved in seawater. It is generally expressed in parts per thousand.
Crab	The term crab is often applied to several different groups of short (nose to tail) decapod crustaceans with thick exoskeletons, but only members of the Brachyura are true.
Stimulus	Stimulus in a nervous system, a factor that triggers sensory transduction.
Antennae	Antennae are paired appendages connected to the front-most segments of arthropods. In crustaceans, they are biramous and present on the first two segments of the head, with the smaller pair known as antennules. In insects, olfactory receptors on the antennae bind to odor molecules, including pheromones.
Habitat	Habitat refers to a place where an organism lives; an environmental situation in which an organism lives.
Mechanoreceptor	A mechanoreceptor is a sensory receptor that responds to mechanical pressure or distortion. There are four main types: Pacinian corpuscles, Meissner's corpuscles, Merkel's discs, and the tympanic membrane.
Carnivore	An animal that eats a diet consisting solely of meat is referred to as a carnivore.
Substrate	A substrate is a molecule which is acted upon by an enzyme. Each enzyme recognizes only the specific substrate of the reaction it catalyzes. A surface in or on which an organism lives.
Transducer	Transducer refers to a device that converts signals from one form to another. A sensory receptor is a transducer that converts environmental stimuli, such as heat, light, or vibration, into electrical signals recognized by the nervous system.
Ventral	The surface or side of the body normally oriented upwards, away from the pull of gravity, is the dorsal side; the opposite side, typically the one closest to the ground when walking on all legs, swimming or flying, is the ventral side.
Soft coral	Soft coral is a group of coral species that are thick and fleshy, and that move and wave in ocean water currents. These corals do not produce calcium carbonate skeleton; they include leather coral, cauliflower coral, and tree coral.
Cnidaria	Cnidaria is a phylum containing some 10,000 species of relatively simple animals found exclusively in aquatic environments (most species are marine). They get their name from cnidocytes, which are specialized cells that carry stinging organelles.
Neuron	The neuron is a major class of cells in the nervous system. In vertebrates, they are found in the brain, the spinal cord and in the nerves and ganglia of the peripheral nervous system, and their primary role is to process and transmit neural information.
Phylum	Phylum is the second highest taxonomic classification for the kingdom Animalia (animals), between kingdom level and class level.
Nerve	A nerve is an enclosed, cable-like bundle of nerve fibers or axons, which includes the glia that ensheath the axons in myelin.

Chemosensor	A Chemosensor is a cell or group of cells that transduce a chemical signal into an action potential. Or, more generally, a chemosensor detects certain chemical stimuli in the environment. There are two main classes of the chemosensor: direct and distance.
Water column	The vertical column of seawater that extends from the surface to the bottom is called water column.
Organ	Organ refers to a structure consisting of several tissues adapted as a group to perform specific functions.
Morphology	The scientific study of organic form, including both its development and function is morphology.
Physiology	The study of the function of cells, tissues, and organs is referred to as physiology.
Ventilation	Ventilation refers to a mechanism that provides contact between an animal's respiratory surface and the air or water to which it is exposed. It is also called breathing.
Predation	Interaction in which one organism uses another, called the prey, as a food source is referred to as predation.
Microorganism	A microorganism is an organism that is so small that it is microscopic (invisible to the naked eye). They are often illustrated using single-celled, or unicellular organisms; however, some unicellular protists are visible to the naked eye, and some multicellular species are microscopic.

Ocean current	An ocean current is any more or less permanent or continuous, directed movement of ocean water that flows in one of the Earth's oceans.
Water column	The vertical column of seawater that extends from the surface to the bottom is called water column.
Plankton	Plankton are drifting organisms that inhabit the water column of oceans, seas, and bodies of fresh water. Plankton abundance and distribution are strongly dependent on factors such as ambient nutrients concentrations, the physical state of the water column, and the abundance of other plankton.
Phytoplankton	Phytoplankton refers to the autotrophic component of the plankton that drifts in the water column. Phytoplankton obtain energy through a process called photosynthesis, and so must live in the well-lit surface layer (termed the euphotic zone) of an ocean, sea, or lake.
Protist	A protist is a heterogeneous group of living things, comprising those eukaryotes that are neither animals, plants, nor fungi. They are a paraphyletic grade, rather than a natural group, and do not have much in common besides a relatively simple organization
Cell	The cell is the structural and functional unit of all living organisms, and is sometimes called the "building block of life."
Flagella	Flagella are whip-like organelle that many unicellular organisms, and some multicellular ones, use to move about.
Organ	Organ refers to a structure consisting of several tissues adapted as a group to perform specific functions.
Zooplankton	Nonphotosynthetic protists that feed on other plankton and are abundant in marine and freshwater environments are zooplankton.
Vertebrate	Vertebrate is a subphylum of chordates, specifically, those with backbones or spinal columns. They started to evolve about 530 million years ago during the Cambrian explosion, which is part of the Cambrian period.
Meroplankton	Meroplankton are organisms that are planktonic for only a part of their life cycles, usually the larval stage. Examples of meroplankton include the larvae of sea urchins, starfish, crustaceans, marine worms, and most fish.
Holoplankton	Holoplankton are organisms that are planktonic for their entire life cycle. Examples of holoplankton include diatoms, radiolarians, dinoflagellates, foraminifera, amphipods, krill, copepods, salps, and jellyfish.
Neuston	Neuston refers to small- to medium-size organisms that live on the surface of the sea.
Surface water	Includes all bodies of water, lakes, rivers, ponds, and so on that are on the surface of the earth, in contrast to groundwater, which lies below the surface is surface water.
Sunlight	Sunlight in the broad sense is the total spectrum of electromagnetic radiation given off by the Sun.
Distribution	Distribution in pharmacology is a branch of pharmacokinetics describing reversible transfer of drug from one location to another within the body.
Crustacean	Member of a group of marine arthropods that contains among others shrimps, crabs, crayfish, and lobsters is called crustacean.
Colony	Colony refers to a cluster or assemblage of microorganisms growing on the surface of an agar culture medium. A colony also refers to several individual organisms of the same species living closely together, usually for mutual benefit, such as stronger defences, the ability to attack bigger prey, etc.

Species	Group of similarly constructed organisms capable of interbreeding and producing fertile offspring is a species.
Dinoflagellate	A unicellular photosynthetic alga with two flagella situated in perpendicular grooves in cellulose plates covering the cell is a dinoflagellate.
Nitrogen	A colorless and tasteless and mostly inert diatomic non-metal gas that is an essential constituent of proteins is nitrogen.
Bacteria	The domain that contains procaryotic cells with primarily diacyl glycerol diesters in their membranes and with bacterial rRNA. Bacteria also is a general term for organisms that are composed of procaryotic cells and are not multicellular.
Ion	Ion refers to an atom or molecule that has gained or lost one or more electrons, thus acquiring an electrical charge.
Magnesium	Magnesium is the chemical element in the periodic table that has the symbol Mg and atomic number 12 and an atomic mass of 24.31.
Calcium	Calcium is the chemical element in the periodic table that has the symbol Ca and atomic number 20. Calcium is a soft grey alkaline earth metal that is used as a reducing agent in the extraction of thorium, zirconium and uranium. Calcium is also the fifth most abundant element in the Earth's crust.
Sodium	Sodium is the chemical element in the periodic table that has the symbol Na (Natrium in Latin) and atomic number 11. Sodium is a soft, waxy, silvery reactive metal belonging to the alkali metals that is abundant in natural compounds (especially halite). It is highly reactive.
Secretion	Secretion is the process of segregating, elaborating, and releasing chemicals from a cell, or a secreted chemical substance or amount of substance.
Cephalopod	Cephalopod refers to a member of a group of mollusks that includes squids and octopuses, characterized by bilateral body symmetry.
Skeleton	In biology, the skeleton or skeletal system is the biological system providing physical support in living organisms.
Bladder	A hollow muscular storage organ for storing urine is a bladder.
Jellyfish	Jellyfish are marine invertebrates belonging to the Scyphozoa class, and in turn the phylum Cnidaria. The body of an adult jellyfish is composed of a bell-shaped, jellylike substance enclosing its internal structure, from which the creature's tentacles suspend.
Copepods	A group of small crustaceans that are important members of marine and freshwater plankton communities is copepods.
Copepod	The group of small crustaceans that is an important member of marine and freshwater plankton communities is copepod. They are very important food organisms for small fish, whales, and other crustaceans such as krill in the ocean and in fresh water.
Receptor	A receptor is a protein on the cell membrane or within the cytoplasm or cell nucleus that binds to a specific molecule (a ligand), such as a neurotransmitter, hormone, or other substance, and initiates the cellular response to the ligand. Receptor, in immunology, the region of an antibody which shows recognition of an antigen.
Predator	A predator is an animal or other organism that hunts and kills other organisms for food in an act called predation.
Pteropod	A mollusc having the anterior lobes of the foot expanded into broad thin winglike swimming organs is a pteropod.

Snail	The name snail applies to most members of the molluscan class Gastropoda that have coiled shells. Other gastropods, which lack a conspicuous shell, are commonly called slugs, and are scattered throughout groups that primarily include snails.
Muscle	Muscle is a contractile form of tissue. It is one of the four major tissue types, the other three being epithelium, connective tissue and nervous tissue. Muscle contraction is used to move parts of the body, as well as to move substances within the body.
Pulse	The rhythmic stretching of the arteries caused by the pressure of blood forced through the arteries by contractions of the ventricles during systole is a pulse.
Arrow worms	Chaetognatha is a phylum of predatory marine worms that are a major component of plankton worldwide. They show some preference for warmer waters. Chaetognaths are transparent and are torpedo shaped sometimes with arrowhead like opaque structures in their heads. They range in size from 3mm to 12cm. The common term for the phylum is arrow worms.
Boundary layer	Boundary layer refers to a layer of still or slow-moving water or air close to the surface of an object.
Temperate	In virology, temperate refers to the life cycle a phage is able to perform. The phage usually integrates its genome into its host chromosome, becoming a prophage.
Phylum	Phylum is the second highest taxonomic classification for the kingdom Animalia (animals), between kingdom level and class level.
Diatom	Diatom refers to a unicellular photosynthetic alga with a unique, glassy cell wall containing silica.
Cell wall	Cell wall refers to a protective layer external to the plasma membrane in plant cells, bacteria, fungi, and some protists; protects the cell and helps maintain its shape.
Silica	Silica is a mineral similar to glass that is the major component of the cell wall, shell, or skeleton of many marine organisms.
Coccolithophore	A coccolithophore is a single-celled algae, or phytoplankton, belonging to the haptophytes. It is distinguished by special calcium carbonate plates (or scales) of unknown purpose called coccoliths.
Spine	A spine is a rigid, pointed surface protuberance or needle-like structure on an animal, shell, or plant, presumably serving as a defense against attack by predators.
Sediment	Sediment is any particulate matter that can be transported by fluid flow and which eventually is deposited as a layer of solid particles on the bed or bottom of a body of water or other liquid.
Cell division	Cell division is the process by which a cell (called the parent cell) divides into two cells (called daughter cells). Cell division is usually a small segment of a larger cell cycle. In meiosis, however, a cell is permanently transformed and cannot divide again.
Daughter cell	A cell formed by cell division of a parent cell is a daughter cell.
Threshold	Electrical potential level at which an action potential or nerve impulse is produced is called threshold.
Gamete	A gamete is a specialized germ cell that unites with another gamete during fertilization in organisms that reproduce sexually. They are haploid cells; that is, they contain one complete set of chromosomes. When they unite they form a zygote—a cell having two complete sets of chromosomes and therefore diploid.
Auxospore	The resistant stage of diatoms that restores the maximum size characteristic to the species is referred to as the auxospore.

Spore	Spore refers to a differentiated, specialized form that can be used for dissemination, for survival of adverse conditions because of its heat and dessication resistance, and/or for reproduction. They are usually unicellular and may develop into vegetative organisms or gametes. A spore may be produced asexually or sexually and are of many types.
Reproduction	Biological reproduction is the biological process by which new individual organisms are produced. Reproduction is a fundamental feature of all known life; each individual organism exists as the result of reproduction by an antecedent.
Population	Group of organisms of the same species occupying a certain area and sharing a common gene pool is referred to as population.
Unicellular	Unicellular organisms carry out all the functions of life. Unicellular species are those whose members consist of a single cell throughout their life cycle. This latter qualification is significant since most multicellular organisms consist of a single cell at the beginning of their life cycles.
Boreal	Boreal is usually applied to ecosystems localized to subarctic (Northern hemisphere) and subantarctic (Southern hemisphere) zones, although Austral is also used for the latter. A boreal forest, also known as the taiga, is the set of forest ecosystems that can survive in northern, specifically subarctic, regions.
Cellulose	A large polysaccharide composed of many glucose monomers linked into cable-like fibrils that provide structural support in plant cell walls is referred to as cellulose.
Theca	In botany, the theca of an angiosperm is half of the anther. An anther and its filament form together a typical (or filantherous) stamen, part of the male floral organ.
Binary fission	A means of asexual reproduction in which a parent organism, often a single cell, divides into two individuals of about equal size is called binary fission.
Sexual reproduction	The propagation of organisms involving the union of gametes from two parents is sexual reproduction.
Cyst	A cyst is a closed sac having a distinct membrane and developing abnormally in a cavity or structure of the body. They may occur as a result of a developmental error in the embryo during pregnancy or they may be caused by infections.
Red tide	Red tide is estuarine or marine algal bloom. Red tide is caused by species of dinoflagellates, often present in sufficient numbers (thousands or millions of cells per milliliter) to turn the water red or brown.
Toxin	Toxin refers to a microbial product or component that can injure another cell or organism at low concentrations. Often the term refers to a poisonous protein, but toxins may be lipids and other substances.
Cardiac arrest	A cardiac arrest is the ceszation of normal circulation of the blood due to failure of the ventricles of the heart to contract effectively during systole.
Respiration	Respiration is the process by which an organism obtains energy by reacting oxygen with glucose to give water, carbon dioxide and ATP (energy). Respiration takes place on a cellular level in the mitochondria of the cells and provide the cells with energy.
Bivalve	Bivalve refers to a member of a group of mollusks that includes clams, mussels, scallops, and oysters. They typically have two-part shells, with both parts being more or less symmetrical.
Silicoflagel-ates	Unicellular and eukatyotic members of the phytoplankton that have a star-shaped silica are silicoflagellates.
Chloroplast	A chloroplast is an organelle found in plant cells and eukaryotic algae which conduct photosynthesis. They are similar to mitochondria but are found only in plants. They are

	surrounded by a double membrane with an intermembrane space; they have their own DNA and are involved in energy metabolism;
Cyanobacteria	Cyanobacteria are a phylum of bacteria that obtain their energy through photosynthesis. They are often referred to as blue-green algae, even though it is now known that they are not directly related to any of the other algal groups, which are all eukaryotes.
Algae	The algae consist of several different groups of living organisms that capture light energy through photosynthesis, converting inorganic substances into simple sugars with the captured energy.
Gyre	Gyre refers to a roughly circular pattern of ocean currents, formed because continents interrupt the currents' flow; rotates clockwise in the Northern Hemisphere and counterclockwise in the Southern Hemisphere.
Nitrogen fixation	Nitrogen fixation is the process by which nitrogen is taken from its relatively inert molecular form (N_2) in the atmosphere and converted into nitrogen compounds useful for other chemical processes.
Amino acid	An amino acid is any molecule that contains both amino and carboxylic acid functional groups. They are the basic structural building units of proteins. They form short polymer chains called peptides or polypeptides which in turn form structures called proteins.
Protein	A protein is a complex, high-molecular-weight organic compound that consists of amino acids joined by peptide bonds. They are essential to the structure and function of all living cells and viruses. Many are enzymes or subunits of enzymes.
Chlorophyceae	The Chlorophyceae are one of the classes of green algae, distinguished mainly on the basis of ultrastructural morphology.
Chlorophyta	The Chlorophyta sensu stricto or chlorophyte algae, comprises most of what are commonly called green algae and includes most members of the grade of putatively ancestral scaly flagellates in Prasinophyceae plus members of Ulvophyceae, Trebouxiophyceae, and Chlorophyceae.
Green algae	Green algae refers to photosynthetic protists that include L.nicellular, colonial, and multicellular species with grass green chloroplasts; closely related to true plants.
Lagoon	A lagoon is a body of comparatively shallow salt water separated from the deeper sea by a shallow or exposed sandbank, coral reef, or similar feature. Thus, the enclosed body of water behind a barrier reef or barrier islands or enclosed by an atoll reef is called a lagoon.
Pollution	Any environmental change that adversely affects the lives and health of living things is referred to as pollution.
Benthos	Benthos are the organisms and habitats of the sea floor; in freshwater biology they are the organisms and habitats of the bottoms of lakes, rivers, and creeks.
Seaweed	Phycologists consider seaweed to refer any of a large number of marine benthic algae that are multicellular, macrothallic (large-bodied), and thus differentiated from most algae that tend to be microscopic in size.
Cryptomonad	A cryptomonad is a small group of flagellates, most of which have chloroplasts. They are common in freshwater, and also occur in marine and brackish habitats. Each cell is around 10-50 ìm in size and flattened in shape, with an anterior groove or pocket.
Flagellate	A flagellate is a cell with one or more whip-like organelles called flagella. Some cells in animals may be flagellate, for instance the sperm cells of most phyla. In protists and microscopic animals, flagella are generally used for propulsion.
Segmentation	Segmentation in biology refers to the division of some metazoan bodies and plant body plans

	into a series of semi-repetitive segments, and the question of the benefits and costs of doing so.
Arthropod	Arthropod refers to member of a phylum of invertebrates that contains among other groups crustaceans and insects that have an exoskeleton and jointed appendages.
Maxillae	The maxillae are the largest bones of the face, except for the mandible, and form, by their union, the whole of the upper jaw. They hold the upper teeth, and connect on the left and right to the zygomatic bones (cheek bones). Each assists in forming the boundaries of three cavities, namely, the roof of the mouth, the floor and lateral wall of the nose and the floor of the orbit.
Mandible	The mandible is the largest and strongest bone of the face. It forms the lower jaw and holds the lower teeth in place. In insects, the mandible is the segment forming the mouth.
Antennae	Antennae are paired appendages connected to the front-most segments of arthropods. In crustaceans, they are biramous and present on the first two segments of the head, with the smaller pair known as antennules. In insects, olfactory receptors on the antennae bind to odor molecules, including pheromones.
Chitin	Chitin refers to strong but flexible nitrogenous polysaccharide found in fungal cells and in the exoskeleton of arthropods.
Eye	An eye is an organ that detects light. Different kinds of light-sensitive organs are found in a variety of creatures.
Mammal	Homeothermic vertebrate characterized especially by the presence of hair and mammary glands is a mammal.
Abdomen	The abdomen is a part of the body. In humans, and in many other vertebrates, it is the region between the thorax and the pelvis. In fully developed insects, the abdomen is the third (or posterior) segment, after the head and thorax.
Thorax	The thorax is a division of an animal's body that lies between the head and the abdomen. In humans, the thorax is the region of the body that extends from the neck to the diaphragm, not including the upper limbs.
Stroke	A stroke or cerebrovascular accident (CVA) occurs when the blood supply to a part of the brain is suddenly interrupted.
Compound eye	Compound eye refers to a type of eye, found in arthropods, that is composed of numerous independent subunits called ommatidia. Each ommatidium apparently contributes a piece of a mosaiclike image perceived by the animal.
Genus	In biology, a genus is a taxonomic grouping. That is, in the classification of living organisms, a genus is considered to be distinct from other such genera. A genus has one or more species: if it has more than one species these are likely to be morphologically more similar than species belonging to different genera.
Egg	An egg is the zygote, resulting from fertilization of the ovum. It nourishes and protects the embryo.
Chemosensor	A Chemosensor is a cell or group of cells that transduce a chemical signal into an action potential. Or, more generally, a chemosensor detects certain chemical stimuli in the environment. There are two main classes of the chemosensor: direct and distance.
Radiolarian	Radiolarian are amoeboid protozoa that produce intricate mineral skeletons, typically with a central capsule dividing the cell into inner and outer portions, called endoplasm and ectoplasm. They are found as plankton throughout the ocean, and their shells are important fossils found from the Cambrian onwards.

Scyphozoan	Scyphozoan of the phylum Cnidaria is any of various usually free-swimming marine coelenterates having a gelatinous medusoid stage as the dominant phase of its life cycle
Baleen whales	The distinguishing feature between this suborder and the toothed whales is that baleen whales have baleen plates instead of teeth in the upper jaw, which enables them to filter food from water.
Cephalothorax	The anterior portion of the body of many arachnid and malacostracan arthropods, which consists of the head fused with the trunk is a cephalothorax.
Spermatophore	In a variation on internal fertilization in some animals, the males package their sperm in a container that can be inserted into the female reproductive tract that is called a spermatophore.
Ventral	The surface or side of the body normally oriented upwards, away from the pull of gravity, is the dorsal side; the opposite side, typically the one closest to the ground when walking on all legs, swimming or flying, is the ventral side.
Sperm	Sperm refers to the male sex cell with three distinct parts at maturity: head, middle piece, and tail.
Photophore	An organ that produces bioluminescence is called a photophore.
Amphipods	Amphipods include over 7000 described species of small, shrimp-like crustaceans. Most amphipods are marine, although a few live in freshwater or are terrestrial. One species common to mid-Atlantic and estuarine waters is Jassa falcata.
Cnidaria	Cnidaria is a phylum containing some 10,000 species of relatively simple animals found exclusively in aquatic environments (most species are marine). They get their name from cnidocytes, which are specialized cells that carry stinging organelles.
Nematocyst	In cnidarians, a capsule that contains a threadlike fiber whose release aids in the capture of prey is referred to as nematocyst.
Morphology	The scientific study of organic form, including both its development and function is morphology.
Pneumatophore	Pneumatophore refers to spongy root extending above the surface of the water, produced by a plant growing in water to facilitate oxygen absorption.
Tentacle	An elongate, extensible projection of the body of cnidarians and cephalopod mollusks that may be used for grasping, stinging, and immobilizing prey, and for locomotion is called tentacle.
Commensal	Living on or within another organism without injuring or benefiting the other organism is referred to as commensal.
Host	Host is an organism that harbors a parasite, mutual partner, or commensal partner; or a cell infected by a virus.
Cilia	Numerous short, hairlike structures projecting from the cell surface that enable locomotion are cilia.
Predation	Interaction in which one organism uses another, called the prey, as a food source is referred to as predation.
Embryo	Embryo refers to a developing stage of a multicellular organism. In humans, the stage in the development of offspring from the first division of the zygote until body structures begin to appear, about the ninth week of gestation.
Larva	A free-living, sexually immature form in some animal life cycles that may differ from the adult in morphology, nutrition, and habitat is called larva.

Bulb	An underground food-storage organ that is essentially a modified bud consisting of fleshy leaves that surround and are attached to a small stem is called bulb.
Thaliacea	The Thaliacea comprise a class of marine animals within the subphylum Urochordata. Unlike their bottom-dwelling relatives the ascidians, thaliaceans are free-floating for their entire lifespan.
Salp	A salp is a primitive, barrel-shaped, free-floating tunicate that move by pumping water through its gelatinous bodies by means of contraction. They are of the class Thaliacea with the other tunicates.
Sea squirts	Tunicates with a sac-like, attached body called a tunic as adults are referred to as sea squirts.
Siphon	The tube-like extension through which water flows in and out of the mantle cavity in bivalves and cephalopods, and in tunicates is called siphon.
Particulate matter	In air pollution, solid particles and liquid droplets, as opposed to material uniformly dispersed among air molecules is particulate matter.
Mucus	Mucus is a slippery secretion of the lining of various membranes in the body (mucous membranes). Mucus aids in the protection of the lungs by trapping foreign particles that enter the nose during normal breathing. Additionally, it prevents tissues from drying out.
Tunicate	One of a group of invertebrate chordates in the subphylum of saclike filter feeders with input and output siphons is referred to as a tunicate.
Fiber	Fiber is a class of materials that are continuous filaments or are in discrete elongated pieces, similar to lengths of thread. They are of great importance in the biology of both plants and animals, for holding tissues together.
Pharynx	The pharynx is the part of the digestive system and respiratory system of many animals immediately behind the mouth and in front of the esophagus.
Water mass	A body of water that can be identified by its temperature and salinity is water mass.
Food chain	Food chain refers to a sequence of food transfers from one species to another within an ecosystem.
Ciliate	Ciliate is one of the most important groups of protists, common almost everywhere there is water. They are a type of protozoan that moves by means of cilia.
Food vacuole	The simplest type of digestive cavity, found in single-celled organisms is a food vacuole.
Foraminifera	The Foraminifera, or forams for short, are a large group of amoeboid protists with reticulating pseudopods, fine strands that branch and merge to form a dynamic net. They typically produce a shell, or test, which can have either one or multiple chambers, some becoming quite elaborate in structure.
Micrograph	A micrograph is a photograph or similar image taken through a microscope or similar device to show a magnified image of an item.
Electron	The electron is a light fundamental subatomic particle that carries a negative electric charge. The electron is a spin-1/2 lepton, does not participate in strong interactions and has no substructure.
Calcium carbonate	A mineral that is the major component of the shell, skeleton, and other parts of many organisms are called calcium carbonate.
Cytoplasm	Cytoplasm refers to everything inside a cell between the plasma membrane and the nucleus; consists of a semifluid medium and organelles.

Fission	A means of asexual reproduction whereby a parent separates into two or more genetically identical individuals of about equal size is referred to as fission.
Zygote	Diploid cell formed by the union of sperm and egg is referred to as zygote.
Symbiont	The smaller participant in a symbiotic relationship, living in or on the host is a symbiont.
Capsule	A sticky layer that surrounds the bacterial cell wall, protects the cell surface, and sometimes helps glue the cell to surfaces is called the capsule. In botany, a capsule is a type of dry fruit as in the poppy, iris, foxglove, etc. as well as another term for the sporangium of mosses and hornworts.
Zooxanthellae	Zooxanthellae refers to dinoflagellates and other algae that live as symbionts within coral.
Nutrition	Nutrition refers to collectively, the processes involved in taking in, assimilating, and utilizing nutrients.
Radiolarian ooze	Radiolarian ooze occurrs in many places in very deep water it is composed mainly of the siliceous skeletons of radiolarians, and calcareous matter that's dissolved by the lage percentage of carbon dioxide in the water at these depths.
Deep scattering layer	A sound reflecting layer of many types of organisms within the ocean that migrates daily from the mesopelagic zone to the epipelagic zone is referred to as deep scattering layer.
Migration	Migration occurs when living things move from one biome to another. In most cases organisms migrate to avoid local shortages of food, usually caused by winter. Animals may also migrate to a certain location to breed, as is the case with some fish.
Chlorophyll	Chlorophyll is a green photosynthetic pigment found in plants, algae, and cyanobacteria. In plant photosynthesis incoming light is absorbed by chlorophyll and other accessory pigments in the antenna complexes of photosystem I and photosystem II.
Kinetic energy	Kinetic energy refers to energy that is actually doing work; the energy of a mass of matter that is moving. Moving matter performs work by transferring its motion to other matter, such as leg muscles pushing bicycle pedals.
Dispersal	Biological dispersal refers to those processes by which a species maintains or expands the distribution of a population. Dispersal implies movement—movement away from an existing population or away from the parent organisms.
Downstream	Downstream refers to a relative position in DNA or RNA. Relative to the position on the strand, downstream is the region towards the 3' end of the strand.
Estuary	An estuary is a semi-enclosed coastal body of water which has a free connection with the open sea and within which sea water mixes with fresh water. The key feature of an estuary is that it is a mixing place for sea water and a stream or river to supply fresh water.
Sonar	A technique or equipment used to locate objects underwater by the detection of echoes is referred to as sonar.
Diurnal	Diurnal is a term that refers to being active in the daytime.
Ctenophore	A ctenophore is a jellyfish-like animal commonly called a comb jelly. Despite their superficial similarities with jellyfish, they have a much more complex anatomy.
Annelid	Annelid refers to member of a phylum of invertebrates that contains segmented worms, such as the earthworm and the clam worm.
Biological clock	Biological clock refers to an internal timekeeper that controls an organism's biological rhythms; marks time with or without environmental cues but often requires signals from the environment to remain tuned to an appropriate period.

Evolution	In biology, evolution is the process by which novel traits arise in populations and are passed on from generation to generation. Its action over large stretches of time explains the origin of new species and ultimately the vast diversity of the biological world.
Metabolic rate	Energy expended by the body per unit time is called metabolic rate.
Gradient	Gradient refers to a difference in concentration, pressure, or electrical charge between two regions.
Poikilotherm	An animal whose body temperature tends to vary with the surrounding environment is a poikilotherm. They are cold blooded.
Growth rate	A measure of the change in population size per individual per unit of time is called growth rate.
Adaptation	A biological adaptation is an anatomical structure, physiological process or behavioral trait of an organism that has evolved over a period of time by the process of natural selection such that it increases the expected long-term reproductive success of the organism.
Relative abundance	Differences in the abundance of species within a community is relative abundance.
Bioluminescence	Bioluminescence is the production and emission of light by a living organism as the result of a chemical reaction during which chemical energy is converted to light energy.
Grazing	Grazing refers to consumption of a portion of a plant's tissues, generallyapplied to grasses and other herbaceous vegetation; removing plant or algal cover from a substrate, generally applied to aquatic systems.
Antibiotic	Antibiotic refers to substance such as penicillin or streptomycin that is toxic to microorganisms. Usually a product of a particular microorvanism or plant.
Acid	An acid is a water-soluble, sour-tasting chemical compound that when dissolved in water, gives a solution with a pH of less than 7.
Digestion	Digestion refers to the mechanical and chemical breakdown of food into molecules small enough for the body to absorb; the second main stage of food processing, following ingestion.
Fossil	A preserved remnant or impression of an organism that lived in the past is referred to as fossil.
Herbivore	A herbivore is an animal that is adapted to eat primarily plant matter
Biology	Biology is the branch of science dealing with the study of life. It is concerned with the characteristics, classification, and behaviors of organisms, how species come into existence, and the interactions they have with each other and with the environment.
Bathypelagic	Bathypelagic is from 1000 m down to around 4000 m, by this depth the ocean is almost entirely dark. There are no living plants, and most animals survive by consuming the snow of detritus falling from the zones above, or by preying upon others. Giant squid live at this depth, and here they are hunted by deep-diving sperm whales.
Oxygen	Oxygen is a chemical element in the periodic table. It has the symbol O and atomic number 8. Oxygen is the second most common element on Earth, composing around 46% of the mass of Earth's crust and 28% of the mass of Earth as a whole, and is the third most common element in the universe.
Upwelling	Upward movement of deep, nutrient-rich water along coasts is referred to as upwelling.
Habitat	Habitat refers to a place where an organism lives; an environmental situation in which an organism lives.

Water column	The vertical column of seawater that extends from the surface to the bottom is called water column.
Nekton	Animals, such as fish, that can swim against currents of water are referred to as nekton.
Cephalopod	Cephalopod refers to a member of a group of mollusks that includes squids and octopuses, characterized by bilateral body symmetry.
Reptile	Member of a class of terrestrial vertebrates with internal fertilization, scaly skin, and an egg with a leathery shell is called reptile.
Mammal	Homeothermic vertebrate characterized especially by the presence of hair and mammary glands is a mammal.
Plankton	Plankton are drifting organisms that inhabit the water column of oceans, seas, and bodies of fresh water. Plankton abundance and distribution are strongly dependent on factors such as ambient nutrients concentrations, the physical state of the water column, and the abundance of other plankton.
Boundary layer	Boundary layer refers to a layer of still or slow-moving water or air close to the surface of an object.
Nervous system	The nervous system of an animal coordinates the activity of the muscles, monitors the organs, constructs and processes input from the senses, and initiates actions.
Phylum	Phylum is the second highest taxonomic classification for the kingdom Animalia (animals), between kingdom level and class level.
Siphon	The tube-like extension through which water flows in and out of the mantle cavity in bivalves and cephalopods, and in tunicates is called siphon.
Mantle	The mantle is an organ found in mollusks. It is the dorsal body wall covering the main body, or visceral mass. The epidermis of this organ secretes calcium carbonate to create a shell.
Vertebrate	Vertebrate is a subphylum of chordates, specifically, those with backbones or spinal columns. They started to evolve about 530 million years ago during the Cambrian explosion, which is part of the Cambrian period.
Eye	An eye is an organ that detects light. Different kinds of light-sensitive organs are found in a variety of creatures.
Invertebrate	Invertebrate is a term coined by Jean-Baptiste Lamarck to describe any animal without a spinal column. It therefore includes all animals except vertebrates (fish, reptiles, amphibians, birds and mammals).
Tentacle	An elongate, extensible projection of the body of cnidarians and cephalopod mollusks that may be used for grasping, stinging, and immobilizing prey, and for locomotion is called tentacle.
Gland	A gland is an organ in an animal's body that synthesizes a substance for release such as hormones, often into the bloodstream or into cavities inside the body or its outer surface.
Olfaction	Olfaction, the sense of odor, is the detection of chemicals dissolved in air.In vertebrates smells are sensed by the olfactory epithelium located in the nasal cavity and first processed by the olfactory bulb in the olfactory system. In insects smells are sensed by sensilia located on the antenna and first processed by the antennal lobe.
Predator	A predator is an animal or other organism that hunts and kills other organisms for food in an act called predation.
Mollusk	The mollusk is the large and diverse phylum Mollusca, which includes a variety of familiar creatures well-known for their decorative shells or as seafood. These range from tiny snails, clams, and abalone to the octopus, cuttlefish and squid (which are considered the most

	intelligent invertebrates).
Muscle	Muscle is a contractile form of tissue. It is one of the four major tissue types, the other three being epithelium, connective tissue and nervous tissue. Muscle contraction is used to move parts of the body, as well as to move substances within the body.
Crustacea	The animal phylum that includes mostly aquatic animals such as crabs, lobsters, crayfish, and shrimp is referred to as crustacea.
Cell	The cell is the structural and functional unit of all living organisms, and is sometimes called the "building block of life."
Internal fertilization	Internal fertilization is a form of animal fertilization of an ovum by spermatozoon within the body of an inseminated animal, whether female or hermaphroditic.
Spermatophore	In a variation on internal fertilization in some animals, the males package their sperm in a container that can be inserted into the female reproductive tract that is called a spermatophore.
Substratum	Substratum is the underlying layer, or base to which a lichen is fixed.
Egg	An egg is the zygote, resulting from fertilization of the ovum. It nourishes and protects the embryo.
Absorption	Absorption is a physical or chemical phenomenon or a process in which atoms, molecules, or ions enter some bulk phase - gas, liquid or solid material. In nutrition, amino acids are broken down through digestion, which begins in the stomach.
Habitat	Habitat refers to a place where an organism lives; an environmental situation in which an organism lives.
Calcium carbonate	A mineral that is the major component of the shell, skeleton, and other parts of many organisms are called calcium carbonate.
Dorsal	In anatomy, the dorsal is the side in which the backbone is located. This is usually the top of an animal, although in humans it refers to the back.
Salt	Salt is a term used for ionic compounds composed of positively charged cations and negatively charged anions, so that the product is neutral and without a net charge.
Diffusion	Diffusion refers to the spontaneous movement of particles of any kind from where they are more concentrated to where they are less concentrated.
Calcareous	Calcareous refers to a sediment, sedimentary rock, or soil type which is formed from or contains a high proportion of calcium carbonate in the form of calcite or aragonite. It is also used to refer to relatively alkaline soil.
Septa	Septa are thin walls or partitions between the internal chambers (camerae) of the shell of a cephalopod, namely nautiloids or ammonoids.
Tissue	Group of similar cells which perform a common function is called tissue.
Cartilaginous fish	A fish that has a flexible skeleton made of cartilage is a cartilaginous fish.
Bony fish	Osteichthyes are the bony fish, a group paraphyletic to the land vertebrates, which are sometimes included. Most belong to the Actinopterygii.
Submarine canyon	A Submarine canyon is a steep-sided valley on the seafloor of the continental slope. Submarine canyons are generally found as extensions to large rivers, and have been found to extend 1km below sea level, and extend for hundreds of kilometers.
Coral reef	A coral reef is a type of biotic reef that develops in tropical waters by the growth of

	coralline algae, hermatypic corals, and other marine organisms.
Species	Group of similarly constructed organisms capable of interbreeding and producing fertile offspring is a species.
Digestive system	The organ system that ingests food, breaks it down into smaller chemical units, and absorbs the nutrient molecules is referred to as the digestive system.
Spinal cord	The spinal cord is a part of the vertebrate nervous system that is enclosed in and protected by the vertebral column (it passes through the spinal canal). It consists of nerve cells. The spinal cord carries sensory signals and motor innervation to most of the skeletal muscles in the body.
Subphylum	A subphylum is a taxonomic rank intermediate between phylum and superclass.
Skeleton	In biology, the skeleton or skeletal system is the biological system providing physical support in living organisms.
Chordata	Chordata refers to the phylum of the chordates; characterized by a dorsal hollow nerve cord, a notochord, gill structures, and a post-anal tail; includes lancelets, tunicates, and vertebrates.
Chondrichthye	A chondrichthye fish is a jawed fish with paired fins, paired nostrils, scales, two-chambered hearts, and skeletons made of cartilage.
Osteichthyes	Class Osteichthyes are the bony fish, a group paraphyletic to the land vertebrates, which are sometimes included. Most belong to the Actinopterygii. The other seven living species are called lobe-finned fish, and include lungfish and coelacanths.
Cartilage	Cartilage is a type of dense connective tissue. Cartilage is composed of cells called chondrocytes which are dispersed in a firm gel-like ground substance, called the matrix. Cartilage is avascular (contains no blood vessels) and nutrients are diffused through the matrix.
Zooplankton	Nonphotosynthetic protists that feed on other plankton and are abundant in marine and freshwater environments are zooplankton.
Adaptation	A biological adaptation is an anatomical structure, physiological process or behavioral trait of an organism that has evolved over a period of time by the process of natural selection such that it increases the expected long-term reproductive success of the organism.
Ecology	Ecology is the scientific study of the distribution and abundance of living organisms and how these properties are affected by interactions between the organisms and their environment.
Spine	A spine is a rigid, pointed surface protuberance or needle-like structure on an animal, shell, or plant, presumably serving as a defense against attack by predators.
Caudal	Pertaining to the tail, or to the posterior part of the body is called caudal.
Ocean Sunfish	The Ocean Sunfish is a unique pelagic fish, perhaps the most massive bony fish in the world. Specimens of ocean sunfish have been observed up to 3.33 m (11 ft) in length and weighing up to 2,300 kg (5070 lbs).
Element	A chemical element, often called simply element, is a chemical substance that cannot be divided or changed into other chemical substances by any ordinary chemical technique. An element is a class of substances that contain the same number of protons in all its atoms.
Vertebral column	In human anatomy, the vertebral column is a column of vertebrae situated in the dorsal aspect of the abdomen. It houses the spinal cord in its spinal canal.
Collagen	Collagen is the main protein of connective tissue in animals and the most abundant protein in mammals, making up about 1/4 of the total. It is one of the long, fibrous structural proteins

	whose functions are quite different from those of globular proteins such as enzymes.
Helical	A helix is a twisted shape like a spring, screw or a spiral staircase. Helices are important in biology, as DNA is helical and many proteins have helical substructures, known as alpha helices.
Pectoral fins	The pectoral fins are located on each side, usually at a middle height on the body just behind the head of fish.
Ventral	The surface or side of the body normally oriented upwards, away from the pull of gravity, is the dorsal side; the opposite side, typically the one closest to the ground when walking on all legs, swimming or flying, is the ventral side.
Oxygen	Oxygen is a chemical element in the periodic table. It has the symbol O and atomic number 8. Oxygen is the second most common element on Earth, composing around 46% of the mass of Earth's crust and 28% of the mass of Earth as a whole, and is the third most common element in the universe.
Blood	Blood is a circulating tissue composed of fluid plasma and cells. The main function of blood is to supply nutrients (oxygen, glucose) and constitutional elements to tissues and to remove waste products.
Gill	An extension of the body surface of an aquatic animal, specialized for gas exchange and/or suspension feeding is called a gill.
Muscle fiber	Cell with myofibrils containing actin and myosin filaments arranged within sarcomeres is a muscle fiber.
Gill arch	A skeletal structure that supports gill filaments and the blood vessels that supply them is called gill arch.
Filament	The stamen is the male organ of a flower. Each stamen generally has a stalk called the filament, and, on top of the filament, an anther. The filament is a long chain of proteins, such as those found in hair, muscle, or in flagella.
Snail	The name snail applies to most members of the molluscan class Gastropoda that have coiled shells. Other gastropods, which lack a conspicuous shell, are commonly called slugs, and are scattered throughout groups that primarily include snails.
Bivalve	Bivalve refers to a member of a group of mollusks that includes clams, mussels, scallops, and oysters. They typically have two-part shells, with both parts being more or less symmetrical.
Organ	Organ refers to a structure consisting of several tissues adapted as a group to perform specific functions.
Lateral line	In fish, the lateral line is a sense organ used to detect movement in the surrounding water, making so-called "lateral echolocation" possible. They are usually visible as faint lines running lengthwise down each side, from the vicinity of the gill covers to the base of the tail.
Inner ear	The Inner ear consists of the cochlea, where a wave is created by a difference in pressure between the scala vestibuli and the scala tympani. The cochlea has three fluid filled sections. The perilymph fluid in the canals differs from the endolymph fluid in the cochlear duct. The organ of Corti is the sensor of pressure variations.
Fiber	Fiber is a class of materials that are continuous filaments or are in discrete elongated pieces, similar to lengths of thread. They are of great importance in the biology of both plants and animals, for holding tissues together.
Otolith	An otolith is a structure in the saccule or utricle of the inner ear. It is comprized of a combination of a gelatinous matrix and calcium carbonate crystals.

Mechanoreceptor	A mechanoreceptor is a sensory receptor that responds to mechanical pressure or distortion. There are four main types: Pacinian corpuscles, Meissner's corpuscles, Merkel's discs, and the tympanic membrane.
Brain	The part of the central nervous system involved in regulating and controlling body activity and interpreting information from the senses transmitted through the nervous system is referred to as the brain.
Nerve	A nerve is an enclosed, cable-like bundle of nerve fibers or axons, which includes the glia that ensheath the axons in myelin.
Color vision	Ability to detect the color of an object, dependent on three kinds of cone cells is called color vision.
Microhabitat	A microhabitat is the immediate surroundings and other physical factors of an individual plant or animal within its habitat.
Predation	Interaction in which one organism uses another, called the prey, as a food source is referred to as predation.
Sunlight	Sunlight in the broad sense is the total spectrum of electromagnetic radiation given off by the Sun.
Trait	In biology, a trait or character is a genetically inherited feature of an organism: Eye color is a character; brown eyes and blue eyes is a trait.
Vertebrae	Vertebrae are the individual bones that make up the vertebral column (aka spine) - a flexuous and flexible column.
Blood vessel	A blood vessel is a part of the circulatory system and function to transport blood throughout the body. The most important types, arteries and veins, are so termed because they carry blood away from or towards the heart, respectively.
Skin	Skin is an organ of the integumentary system composed of a layer of tissues that protect underlying muscles and organs.
Evolution	In biology, evolution is the process by which novel traits arise in populations and are passed on from generation to generation. Its action over large stretches of time explains the origin of new species and ultimately the vast diversity of the biological world.
Temperate	In virology, temperate refers to the life cycle a phage is able to perform. The phage usually integrates its genome into its host chromosome, becoming a prophage.
Distribution	Distribution in pharmacology is a branch of pharmacokinetics describing reversible transfer of drug from one location to another within the body.
Counterillumination	Counterillumination refers to the the lighting of the underside of an animal with bioluminescence which eliminates its silhouette through "downlighting".
Mesopelagic	The pelagic zone is the part of the open sea or ocean comprising the water column, i.e. all of the sea other than that near the coast or the sea floor. Mesopelagic (from 200 metres down to around 1000 metres) - the twilight zone. Although some light penetrates this deep, it is insufficient for photosynthesis.
Tubular eyes	Tubular eyes refers to specialized eyes of many midwater animals that allow acute upward or downward vision.
Dorsal surface	The dorsal surface is arched from above downward, and is subdivided into two unequal parts by the spine; the portion above the spine is called the supraspinatous fossa, and that below it the infraspinous fossa.
Simultaneous	Simultaneous hermaphrodite is when the organism has both male and female sexual organs at the

hermaphrodite	same time as an adult. Usually, self-fertilization does not occur.
Photophore	An organ that produces bioluminescence is called a photophore.
Digestive tract	The digestive tract is the system of organs within multicellular animals which takes in food, digests it to extract energy and nutrients, and expels the remaining waste.
Abyssopelagic	The pelagic zone is the part of the open sea or ocean comprising the water column, i.e. all of the sea other than that near the coast or the sea floor. Abyssopelagic (from 4000 metres down to above the ocean floor) - no light whatsoever penetrates to this depth, and most creatures are blind and colorless.
Bathypelagic	Bathypelagic is from 1000 m down to around 4000 m, by this depth the ocean is almost entirely dark. There are no living plants, and most animals survive by consuming the snow of detritus falling from the zones above, or by preying upon others. Giant squid live at this depth, and here they are hunted by deep-diving sperm whales.
Fauna	Fauna (animals), a collective term for animal life, as distinct from flora (plant life).
Cetacea	The order Cetacea includes whales, dolphins and porpoises. It is the order of mammals most fully adapted to aquatic life.
Carbon dioxide	Carbon dioxide is an atmospheric gas comprized of one carbon and two oxygen atoms. A very widely known chemical compound, it is frequently called by its formula CO_2. In its solid state, it is commonly known as dry ice.
Mammalia	The class that includes endothermic vertebrates that possess mammary glands and hair is called Mammalia.
Baleen whales	The distinguishing feature between this suborder and the toothed whales is that baleen whales have baleen plates instead of teeth in the upper jaw, which enables them to filter food from water.
Sperm Whale	The Sperm Whale (Physeter macrocephalus) is the largest of all toothed whales and is believed to be the largest toothed animal to ever inhabit Earth.
Vestigial	The remains of body structures that are no longer of adaptive value to the organism and therefore are not maintained by selection is a vestigial.
Countercurrent exchange	The transfer of a substance from a fluid or volume of air moving in one direction to another fluid or volume of air moving in the opposite direction is referred to as countercurrent exchange.
Copulation	Copulation is the union of the external sexual organs of two sexually reproducing animal organisms for insemination and for subsequent internal fertilization, which is fertilization of ova inside organisms.
Gestation	Gestation refers to pregnancy; the state of carrying developing young within the female reproductive tract.
Blowhole	Blowhole refers to the opening on top of a cetacean's head through which the animal breathes, occasionally producing a splash when breathing out.
Breaching	Leaping into the air by whales is referred to as breaching. The act of leaping generates more power than any other act performed by a non-human animal.
Kelp	Kelp are large seaweeds, belonging to the brown algae and classified in the order Laminariales. Despite their appearance they are not grouped with the normal aquatic or land plants (kingdom Plantae), but instead are included in either kingdom Protista or Chromista.
Sponge	An invertebrates that consist of a complex aggregation of cells, including collar cells, and has a skeleton of fibers and/or spicules is a sponge. They are primitive, sessile, mostly

	marine, waterdwelling filter feeders that pump water through their matrix to filter out particulates of food matter.
Jellyfish	Jellyfish are marine invertebrates belonging to the Scyphozoa class, and in turn the phylum Cnidaria. The body of an adult jellyfish is composed of a bell-shaped, jellylike substance enclosing its internal structure, from which the creature's tentacles suspend.
Embryo	Embryo refers to a developing stage of a multicellular organism. In humans, the stage in the development of offspring from the first division of the zygote until body structures begin to appear, about the ninth week of gestation.
Swell	Swell refers to a wave with a flatter, rounded wave crest and trough. Swells are found away from the area where waves are generated by the wind .
Manatee	A Manatee (family Trichechidae, genus Trichechus) is a large aquatic mammal sometimes known as a sea cow. The Trichechidae differ from the Dugongidae in the shape of the skull and the shape of the tail. Their tails are paddle-shaped, while the Dugong's is forked. It is an herbivore, spending most of its time grazing in shallow waters
Colony	Colony refers to a cluster or assemblage of microorganisms growing on the surface of an agar culture medium. A colony also refers to several individual organisms of the same species living closely together, usually for mutual benefit, such as stronger defences, the ability to attack bigger prey, etc.
Migration	Migration occurs when living things move from one biome to another. In most cases organisms migrate to avoid local shortages of food, usually caused by winter. Animals may also migrate to a certain location to breed, as is the case with some fish.
Barnacle	A barnacle is a type of arthropod belonging to infraclass Cirripedia in the subphylum Crustacea and is hence distantly related to crabs and lobsters. They live attached to surfaces and are typically enclosed by heavy calcareous plates are referred to as barnacles.
Gastropod	A member of the largest group of mollusks, including snails and slugs is a gastropod.
Seaweed	Phycologists consider seaweed to refer any of a large number of marine benthic algae that are multicellular, macrothallic (large-bodied), and thus differentiated from most algae that tend to be microscopic in size.
Starfish	Starfish are marine invertebrates belonging to phylum Echinodermata, class Asteroidea. They exhibit a superficially radial symmetry, typically with five or more "arms" which radiate from an indistinct disk. They do not have movable skeletons, but instead possess a hydraulic water vascular system.
Mussel	A mussel is a bivalve mollusk that can be found in lakes, rivers, creeks, intertidal areas, and throughout the ocean. The saltwater mussels (family Mytilidae) and freshwater mussels (family Unionidae) are not thought to be closely related and are grouped in different subclasses, despite considerable similarities in appearance.
Coexistence	Coexistence refers to occurrence of two or more species in the same habitat; usually applied to potentially competing species.
Fission	A means of asexual reproduction whereby a parent separates into two or more genetically identical individuals of about equal size is referred to as fission.
Chiton	A mollusc that has a shell divided into eight overlapping plates is a chiton.
Grazing	Grazing refers to consumption of a portion of a plant's tissues, generallyapplied to grasses and other herbaceous vegetation; removing plant or algal cover from a substrate, generally applied to aquatic systems.
Sea anemone	Named after a terrestrial flower, the sea anemone is a group of water-dwelling, filter

	feeding animals of the order Actinaria. As a cnidarian, it is closely related to coral and jellyfish.
Melon	A fatty structure on the forehead of some cetaceans that is used to direct sound waves emitted during echolocation is called melon. The melon is also the fruit and plant of a typically vine-like herb that was first cultivated more than 4000 years ago (~ 2000 BC) in Persia and Africa.
Population	Group of organisms of the same species occupying a certain area and sharing a common gene pool is referred to as population.
Pinniped	Pinniped refers to mammals with paddle-shaped flippers: seals, eared seals, and the walrus.
Baleen	Baleen is the substance made of keratin that makes up the filtering plates that hang from the upper jaws of baleen whales.
Suspension feeder	An animal that extracts food particles suspended in the surrounding water for food is referred to as a suspension feeder.
Threshold	Electrical potential level at which an action potential or nerve impulse is produced is called threshold.
Rorquals	Rorquals are the largest group of baleen whales, with nine species in two genera. They include the Blue, Fin, Humpback, Sei, and Minke Whales, all members of the family have a series of longitudinal folds of skin running from below the mouth back to the navel (except the Sei Whale, which has shorter grooves).
Carnivore	An animal that eats a diet consisting solely of meat is referred to as a carnivore.
Morphology	The scientific study of organic form, including both its development and function is morphology.
Placental mammal	Member of mammalian subclass characterized by the presence of a placenta during the development of an offspring is referred to as placental mammal.
Homeothermy	Homeothermy refers to the ability to maintain a constant body temperature in the face of a fluctuating environmental temperature; warm-bloodedness.
Fossil	A preserved remnant or impression of an organism that lived in the past is referred to as fossil.
Genus	In biology, a genus is a taxonomic grouping. That is, in the classification of living organisms, a genus is considered to be distinct from other such genera. A genus has one or more species: if it has more than one species these are likely to be morphologically more similar than species belonging to different genera.
Sediment	Sediment is any particulate matter that can be transported by fluid flow and which eventually is deposited as a layer of solid particles on the bed or bottom of a body of water or other liquid.
DNA sequence	A DNA sequence is a succession of letters representing the primary structure of a real or hypothetical DNA molecule or strand, The possible letters are A, C, G, and T, representing the four nucleotide subunits of a DNA strand (adenine, cytosine, guanine, thymine).
DNA sequencing	The process of determining the chemical composition of a DNA molecule is called DNA sequencing.
Sacrum	The sacrum is a large, triangular bone at the base of the vertebral column and at the upper and back part of the pelvic cavity, where it is inserted like a wedge between the two hip bones. Its upper part or base articulates with the last lumbar vertebra, its apex with the coccyx.

Ilium	The ilium is divisible into two parts, the body and the ala; the separation is indicated on the internal surface by a curved line, the arcuate line, and on the external surface by the margin of the acetabulum.
Sequencing	Determining the order of nucleotides in a DNA or RNA molecule or the order of amino acids in a protein is sequencing.
Gene	Gene refers to a discrete unit of hereditary information consisting of a specific nucleotide sequence in DNA . Most of the genes of a eukaryote are located in its chromosomal DNA; a few are carried by the DNA of mitochondria and chloroplasts.
Comparative analysis	An approach to studying evolution in which hypotheses are tested by measuring the distribution of states among a large number of species is called comparative analysis.
Polarity	In cell biology, polarity refers to cells not being point-symmetrical in their spatial organization. In horticulture, polarity refers to the condition in which cuttings grow shoots at the distil end and roots at the proximal end.
Cone	Cone refers to a reproductive structure of gymnosperms that produces pollen in males or eggs in females.
Harem	The term Harem is used for the social organization of certain species, such as those in the Hominidae and Equidae families, in groups of females surrounding a single dominant male.
Sirenia	The Sirenia are fully aquatic, herbivorous mammals that inhabit rivers, estuaries and coastal marine waters. Commonly called manatees, dugongs, or seacows, they have major aquatic adaptations.
Dugong	A Dugong is one of the smallest members of the order Sirenia (which also includes the manatees and Steller's Sea Cow). Adults are generally less than 3 meters long. The name Dugong originated from the Malay language duyung meaning lady of the sea or mermaid.
Lactic acid	Lactic acid accumulates in skeletal muscles during extensive anaerobic exercise, causing temporary muscle pain. Lactic acid is quickly removed from muscles when they resume aerobic metabolism.
Hemoglobin	Hemoglobin is the iron-containing oxygen-transport metalloprotein in the red cells of the blood in mammals and other animals. Hemoglobin transports oxygen from the lungs to the rest of the body, such as to the muscles, where it releases the oxygen load.
Metabolic rate	Energy expended by the body per unit time is called metabolic rate.
Atmosphere	Earth's atmosphere is a layer of gases surrounding the planet Earth and retained by the Earth's gravity. It contains roughly 78% nitrogen and 21% oxygen, with trace amounts of other gases.
Circulatory system	The circulatory system or cardiovascular system is the organ system which circulates blood around the body of most animals.
Joint	A joint (articulation) is the location at which two bones make contact (articulate). They are constructed to both allow movement and provide mechanical support.
Lungs	Lungs are the essential organs of respiration in air-breathing vertebrates. Their principal function is to transport oxygen from the atmosphere into the bloodstream, and to excrete carbon dioxide from the bloodstream into the atmosphere.
Biology	Biology is the branch of science dealing with the study of life. It is concerned with the characteristics, classification, and behaviors of organisms, how species come into existence, and the interactions they have with each other and with the environment.
Promontory	In anatomy, a promontory is a projecting part. Especially the projecting angle of the ventral

	side of the sacrum where it joins the last lumbar vertebra and where there is a prominence on the inner wall of the tympanum of the ear.
Territory	In ethology, sociobiology and behavioral ecology, the term territory refers to any geographical area that an animal of a particular species consistently defends against conspecifics (and, occasionally, animals of other species).
Mandible	The mandible is the largest and strongest bone of the face. It forms the lower jaw and holds the lower teeth in place. In insects, the mandible is the segment forming the mouth.
Colonization	Colonization is the act where life forms move into a distant area where their kind is sparse or not yet existing at all and set up new settlements in the area. Colonization applies to all life forms in a sense though it is most often used in reference to insects and humans.
Tundra	A biome at the northernmost limits of plant growth and at high altitudes, characterized by dwarf woody shrubs, grasses, mosses, and lichens is referred to as tundra.
Crab	The term crab is often applied to several different groups of short (nose to tail) decapod crustaceans with thick exoskeletons, but only members of the Brachyura are true.
Mole	The atomic weight of a substance, expressed in grams. One mole is defined as the mass of 6.0222 3 1023 atoms.
Crustacean	Member of a group of marine arthropods that contains among others shrimps, crabs, crayfish, and lobsters is called crustacean.
Snake	A snake also known as ophidians, are cold blooded legless reptiles closely related to lizards, which share the order Squamata.
Isthmus	An isthmus is a narrow strip of land that is bordered on two sides by water and connects two larger land masses. It is the inverse of a strait (which lies between two land masses and connects two larger bodies of water).
Coral	Group of cnidarians having a calcium carbonate skeleton that participate in the formation of reefs is coral.
Phylogeny	The science that explores the evolutionary relationships among organisms and seeks to reconstruct evolutionary history is phylogeny.
Bioluminescence	Bioluminescence is the production and emission of light by a living organism as the result of a chemical reaction during which chemical energy is converted to light energy.
Physiology	The study of the function of cells, tissues, and organs is referred to as physiology.
Speciation	Speciation refers to the evolutionary process by which new biological species arise. All forms of speciation have actually taken place over the course of evolution, though it still remains a subject of debate as to the relative importance of each mechanism in driving biodiversity.
Demography	Demography refers to the statistical study of population. It encompasses the study of the size, structure and distribution of populations, and how populations change over time due to births, deaths, migration and ageing.
Amphipods	Amphipods include over 7000 described species of small, shrimp-like crustaceans. Most amphipods are marine, although a few live in freshwater or are terrestrial. One species common to mid-Atlantic and estuarine waters is Jassa falcata.
Biochemistry	Biochemistry studies how complex chemical reactions give rise to life. It is a hybrid branch of chemistry which specialises in the chemical processes in living organisms.
Mitochondrial DNA	Mitochondrial DNA is DNA not located in the nucleus of the cell but in the mitochondria, parts of the cell that generate fuel in the form of adenosine triphosphate, which drives the

varied machinery of the cell. Unlike most of the cell, the function of which is defined by the nuclear DNA, the mitochondria have their own DNA and are assumed to have evolved separately.

Insertion At DNA level, an insertion means the insertion of a few base pairs into a genetic sequence. This can often happen in microsatellite regions due to the DNA polymerase slipping.

Population structure Population structure refers to the proportions of individuals in a population belonging to different age classes. Also, the distribution of the population in space.

Zoology Zoology is the biological discipline which involves the study of animals.

Haplotype A haplotype is the genetic constitution of an individual chromosome.

Plankton	Plankton are drifting organisms that inhabit the water column of oceans, seas, and bodies of fresh water. Plankton abundance and distribution are strongly dependent on factors such as ambient nutrients concentrations, the physical state of the water column, and the abundance of other plankton.
Phytoplankton	Phytoplankton refers to the autotrophic component of the plankton that drifts in the water column. Phytoplankton obtain energy through a process called photosynthesis, and so must live in the well-lit surface layer (termed the euphotic zone) of an ocean, sea, or lake.
Population	Group of organisms of the same species occupying a certain area and sharing a common gene pool is referred to as population.
Sunlight	Sunlight in the broad sense is the total spectrum of electromagnetic radiation given off by the Sun.
Diatom	Diatom refers to a unicellular photosynthetic alga with a unique, glassy cell wall containing silica.
Temperate	In virology, temperate refers to the life cycle a phage is able to perform. The phage usually integrates its genome into its host chromosome, becoming a prophage.
Boreal	Boreal is usually applied to ecosystems localized to subarctic (Northern hemisphere) and subantarctic (Southern hemisphere) zones, although Austral is also used for the latter. A boreal forest, also known as the taiga, is the set of forest ecosystems that can survive in northern, specifically subarctic, regions.
Species	Group of similarly constructed organisms capable of interbreeding and producing fertile offspring is a species.
Surface water	Includes all bodies of water, lakes, rivers, ponds, and so on that are on the surface of the earth, in contrast to groundwater, which lies below the surface is surface water.
Dinoflagellate	A unicellular photosynthetic alga with two flagella situated in perpendicular grooves in cellulose plates covering the cell is a dinoflagellate.
Zooplankton	Nonphotosynthetic protists that feed on other plankton and are abundant in marine and freshwater environments are zooplankton.
Nitrate	Nitrate refers to a salt of nitric acid; a compound containing the radical NO_3; biologically, the final form of nitrogen from the oxidation of organic nitrogen compounds.
Copepods	A group of small crustaceans that are important members of marine and freshwater plankton communities is copepods.
Copepod	The group of small crustaceans that is an important member of marine and freshwater plankton communities is copepod. They are very important food organisms for small fish, whales, and other crustaceans such as krill in the ocean and in fresh water.
Invertebrate	Invertebrate is a term coined by Jean-Baptiste Lamarck to describe any animal without a spinal column. It therefore includes all animals except vertebrates (fish, reptiles, amphibians, birds and mammals).
Jellyfish	Jellyfish are marine invertebrates belonging to the Scyphozoa class, and in turn the phylum Cnidaria. The body of an adult jellyfish is composed of a bell-shaped, jellylike substance enclosing its internal structure, from which the creature's tentacles suspend.
Predator	A predator is an animal or other organism that hunts and kills other organisms for food in an act called predation.
Water column	The vertical column of seawater that extends from the surface to the bottom is called water column.

Respiration	Respiration is the process by which an organism obtains energy by reacting oxygen with glucose to give water, carbon dioxide and ATP (energy). Respiration takes place on a cellular level in the mitochondria of the cells and provide the cells with energy.
Cell	The cell is the structural and functional unit of all living organisms, and is sometimes called the "building block of life."
Limiting factor	In biology, agricultural science, physiology, and ecology, a limiting factor is one that controls a process, such as organism growth or species population size or distribution. The availability of food, predation pressure, or availability of shelter are examples of factors that could be limiting for a species population in a specific area.
Photosynthesis	Photosynthesis is a biochemical process in which plants, algae, and some bacteria harness the energy of light to produce food. Ultimately, nearly all living things depend on energy produced from photosynthesis for their nourishment, making it vital to life on Earth.
Oxygen	Oxygen is a chemical element in the periodic table. It has the symbol O and atomic number 8. Oxygen is the second most common element on Earth, composing around 46% of the mass of Earth's crust and 28% of the mass of Earth as a whole, and is the third most common element in the universe.
Absorption	Absorption is a physical or chemical phenomenon or a process in which atoms, molecules, or ions enter some bulk phase - gas, liquid or solid material. In nutrition, amino acids are broken down through digestion, which begins in the stomach.
Extinction	In biology and ecology, extinction is the ceasing of existence of a species or group of taxa. The moment of extinction is generally considered to be the death of the last individual of that species.The death of all members of a species is extinction.
Pigment	Pigment is any material resulting in color in plant or animal cells which is the result of selective absorption.
Vertical mixing	Exchange of water between deep and surface layers is vertical mixing.
Thermocline	The thermocline is a layer within a body of water where the temperature changes rapidly with depth. Because water is not perfectly transparent, almost all sunlight is absorbed in the surface layer, which heats up.
Salinity	Salinity refers to the total amount of salts dissolved in seawater. It is generally expressed in parts per thousand.
Runoff	Water entering rivers, lakes, reservoirs, or the ocean from land surfaces is runoff.
Fjord	A fjord is a narrow inlet of the sea between cliffs or steep slopes, which results from marine inundation of a glaciated valley. Typical characteristics of a fjord include: a narrow inlet, a bottom glacially eroded significantly below sea level, steep-sided walls which continue to descend below the sea surface, greater depths in the upper and middle reaches than on the seaward side, and communication with the open sea.
Detritus	Detritus is organic waste material from decomposing dead plants or animals.
Stratification	Stratification is the process of pretreating seeds to simulate natural conditions that a seed must endure before germination. Many seed species have what is called an embryonic dormancy and generally speaking will not sprout until this dormancy is broken.
Bivalve	Bivalve refers to a member of a group of mollusks that includes clams, mussels, scallops, and oysters. They typically have two-part shells, with both parts being more or less symmetrical.
Excretion	Excretion is the biological process by which an organism chemically separates waste products from its body. The waste products are then usually expelled from the body by elimination.

Primary production	Primary production is the production of organic compounds from inorganic materials principally through the process of photosynthesis.
Watershed	The total land area that drains directly or indirectly into a particular stream or river. The watershed is generally named for the stream or river into which it drains.
Carbon	Carbon is a chemical element in the periodic table that has the symbol C and atomic number 6. An abundant nonmetallic, tetravalent element, carbon has several allotropic forms.
Bacteria	The domain that contains procaryotic cells with primarily diacyl glycerol diesters in their membranes and with bacterial rRNA. Bacteria also is a general term for organisms that are composed of procaryotic cells and are not multicellular.
Estuary	An estuary is a semi-enclosed coastal body of water which has a free connection with the open sea and within which sea water mixes with fresh water. The key feature of an estuary is that it is a mixing place for sea water and a stream or river to supply fresh water.
Carbon dioxide	Carbon dioxide is an atmospheric gas comprized of one carbon and two oxygen atoms. A very widely known chemical compound, it is frequently called by its formula CO_2. In its solid state, it is commonly known as dry ice.
Downstream	Downstream refers to a relative position in DNA or RNA. Relative to the position on the strand, downstream is the region towards the 3' end of the strand.
Phosphorus	Phosphorus is the chemical element in the periodic table that has the symbol P and atomic number 15.
Nitrogen	A colorless and tasteless and mostly inert diatomic non-metal gas that is an essential constituent of proteins is nitrogen.
Chemical energy	Chemical energy refers to energy stored in the chemical bonds of molecules; a form of potential energy.
Transformation	Transformation is the genetic alteration of a cell resulting from the introduction, uptake and expression of foreign genetic material (DNA or RNA).
Light reactions	The first of two stages in photosynthesis, the steps in which solar energy is absorbed and converted to chemical energy in the form of ATP and NADPH. The light reactions power the sugar-producing Calvin cycle but produce no sugar themselves.
Calvin cycle	Calvin cycle refers to the second of two stages of photosynthesis; a cyclic series of chemical reactions that occur in the stroma of a chloroplast, using the carbon in CO_2 and the ATP and NADPH produced by the light reactions to make the energy-rich sugars.
Gross primary productivity	Gross primary productivity refers to production before respiration losses are subtracted; photosynthetic production for plants and metabolizable production for animals.
Point source	A source of a pollutant such as discharges from factory drains or outlets from sewage treatment plants is a point source.
Starch	Biochemically, starch is a combination of two polymeric carbohydrates (polysaccharides) called amylose and amylopectin.
Organelle	In cell biology, an organelle is one of several structures with specialized functions, suspended in the cytoplasm of a eukaryotic cell.
Chloroplast	A chloroplast is an organelle found in plant cells and eukaryotic algae which conduct photosynthesis. They are similar to mitochondria but are found only in plants. They are surrounded by a double membrane with an intermembrane space; they have their own DNA and are involved in energy metabolism;
Carbohydrate	Carbohydrate is a chemical compound that contains oxygen, hydrogen, and carbon atoms. They

	consist of monosaccharide sugars of varying chain lengths and that have the general chemical formula $C_n(H_2O)_n$ or are derivatives of such.
Photosynthetic pigment	A photosynthetic pigment is a pigment that is present in chloroplasts or photosynthetic bacteria and captures the light energy necessary for photosynthesis.
Chlorophyll A	A green pigment in chloroplasts that participates directly in the light reactions is referred to as chlorophyll A. Due to its central role in the reaction center, is essential for photosynthesis.
Electron	The electron is a light fundamental subatomic particle that carries a negative electric charge. The electron is a spin-1/2 lepton, does not participate in strong interactions and has no substructure.
Molecule	A molecule is the smallest particle of a pure chemical substance that still retains its chemical composition and properties.
Enzyme	An enzyme is a protein that catalyzes, or speeds up, a chemical reaction. They are essential to sustain life because most chemical reactions in biological cells would occur too slowly, or would lead to different products, without them.
Acid	An acid is a water-soluble, sour-tasting chemical compound that when dissolved in water, gives a solution with a pH of less than 7.
Ribulose	A 5-carbon sugar molecule used in photosynthesis is ribulose.
Rubisco	RUBISCO is the common name of ribulose-1,5-bisphosphate carboxylase/oxygenase, an enzyme that is used in the Calvin cycle to catalyze the first major step of carbon fixation. RUBISCO catalyzes either the carboxylation or oxygenation of ribulose-1,5-bisphosphate (also known as RuBP) with carbon dioxide or oxygen.
Sugar	A sugar is the simplest molecule that can be identified as a carbohydrate. These include monosaccharides and disaccharides, trisaccharides and the oligosaccharides. The term "glyco-" indicates the presence of a sugar in an otherwise non-carbohydrate substance.
Residence time	Residence time refers to the ratio of the size of a compartment to the flux through it, expressed in units of time; thus, the average time spent by energy or a substance in the compartment.
Regeneration	Regeneration is the ability to restore lost or damaged tissues, organs or limbs. It is a common feature in invertebrates, but far more limited in most vertebrates.
Pulse	The rhythmic stretching of the arteries caused by the pressure of blood forced through the arteries by contractions of the ventricles during systole is a pulse.
Primary productivity	Primary productivity is the amount of light energy converted to chemical energy in a given amount of time. It usually corresponds to the amount of photosynthesis that occurred.
Distribution	Distribution in pharmacology is a branch of pharmacokinetics describing reversible transfer of drug from one location to another within the body.
Attenuation	Regulatory mechanism for controlling frequency of transcription of a gene, which employs a nucleotide sequence that leads to premature transcription is called attenuation.
Wavelength	The distance between crests of adjacent waves, such as those of the electromagnetic spectrum is wavelength.
Radiation	The emission of electromagnetic waves by all objects warmer than absolute zero is referred to as radiation.
Bleaching	Bleaching is a phenomenon where severely stressed corals turn white, frequently due to high water temperatures. This color change is mainly due to the corals expelling their

zooxanthellae.

Accessory pigment An accessory pigment is a photosynthetic pigment other than the chlorophylls that enable an organism to utilize more colors of the visible light spectrum for photosynthesis, e.g., carotinoids phycoerythrins and phycocyanin.

Chlorophyll Chlorophyll is a green photosynthetic pigment found in plants, algae, and cyanobacteria. In plant photosynthesis incoming light is absorbed by chlorophyll and other accessory pigments in the antenna complexes of photosystem I and photosystem II.

Fucoxanthin Fucoxanthin is a carotenoid, with formula $C_{40}H_{60}O_6$. It is found as an accessory pigment in the chloroplasts of brown algae and most other heterokonts, giving them a brown or olive-green color. They absorb red light.

Action spectrum An action spectrum is the rate of a physiological activity plotted against wavelength of light.

Visible light That portion of the electromagnetic spectrum detected as various colors by the human eye, ranging in wavelength from about 380 nm to about 750 rim is called visible light.

Evolution In biology, evolution is the process by which novel traits arise in populations and are passed on from generation to generation. Its action over large stretches of time explains the origin of new species and ultimately the vast diversity of the biological world.

Photorespiration Photorespiration is an alternate pathway for Rubisco, the main enzyme of photosynthesis (specifically, the Calvin cycle). Although Rubisco favors carbon dioxide, it can also use oxygen, producing a glycolate and a glycerate.

Adaptation A biological adaptation is an anatomical structure, physiological process or behavioral trait of an organism that has evolved over a period of time by the process of natural selection such that it increases the expected long-term reproductive success of the organism.

Protist A protist is a heterogeneous group of living things, comprising those eukaryotes that are neither animals, plants, nor fungi. They are a paraphyletic grade, rather than a natural group, and do not have much in common besides a relatively simple organization

Element A chemical element, often called simply element, is a chemical substance that cannot be divided or changed into other chemical substances by any ordinary chemical technique. An element is a class of substances that contain the same number of protons in all its atoms.

Copper Copper is a chemical element in the periodic table that has the symbol Cu (L.: Cuprum) and atomic number 29. It is a ductile metal with excellent electrical conductivity, and finds extensive use as a building material, as an electrical conductor, and as a component of various alloys.

Iron Iron is essential to all organisms, except for a few bacteria. It is mostly stably incorporated in the inside of metalloproteins, because in exposed or in free form it causes production of free radicals that are generally toxic to cells.

Compensation point The light intensity at which the rates of photosynthesis and of cellular respiration are equal is referred to as compensation point.

Heterotrophic Incapable of synthesizing food and therefore dependent on other organisms for it is called heterotrophic.

Nutrition Nutrition refers to collectively, the processes involved in taking in, assimilating, and utilizing nutrients.

Peptide Peptide is the family of molecules formed from the linking, in a defined order, of various amino acids. The link between one amino acid residue and the next is an amide bond, and is sometimes referred to as a peptide bond.

Seaweed	Phycologists consider seaweed to refer any of a large number of marine benthic algae that are multicellular, macrothallic (large-bodied), and thus differentiated from most algae that tend to be microscopic in size.
Marine snow	In the deep ocean, marine snow is a continuous shower of mostly organic detritus falling from the upper layers of the water column. Its origin lies in activities within the productive photic zone.
Macromolecule	A macromolecule is a molecule of high relative molecular mass, the structure of which essentially comprises the multiple repetition of units derived, actually or conceptually, from molecules of low relative molecular mass.
Mucus	Mucus is a slippery secretion of the lining of various membranes in the body (mucous membranes). Mucus aids in the protection of the lungs by trapping foreign particles that enter the nose during normal breathing. Additionally, it prevents tissues from drying out.
Tunicate	One of a group of invertebrate chordates in the subphylum of saclike filter feeders with input and output siphons is referred to as a tunicate.
Surface layer	The upper layer of water that is mixed by wind, waves, and currents is the surface layer .
Photic zone	The region of ocean or lake water where light is strong enough to support photosynthesis is referred to as photic zone.
Biomass	Biomass is organic non-fossil material, collectively. In other words, biomass comprises the mass of all biological organisms, dead or alive, excluding biological mass that has been transformed by geological processes into substances such as coal or petroleum.
Upwelling	Upward movement of deep, nutrient-rich water along coasts is referred to as upwelling.
Fecal	Fecal is the solid excretory wastes of animals. Consists of undigested material passing through the gut and bacteria that have begun to feed on it.
Glucose	Glucose, a simple monosaccharide sugar, is one of the most important carbohydrates and is used as a source of energy in animals and plants. Glucose is one of the main products of photosynthesis and starts respiration.
Protein synthesis	The process whereby the tRNA utilizes the mRNA as a guide to arrange the amino acids in their proper sequence according to the genetic information in the chemical code of DNA is referred to as protein synthesis.
Amino acid	An amino acid is any molecule that contains both amino and carboxylic acid functional groups. They are the basic structural building units of proteins. They form short polymer chains called peptides or polypeptides which in turn form structures called proteins.
Protein	A protein is a complex, high-molecular-weight organic compound that consists of amino acids joined by peptide bonds. They are essential to the structure and function of all living cells and viruses. Many are enzymes or subunits of enzymes.
Urea	Urea is an organic compound of carbon, nitrogen, oxygen and hydrogen, CON_2H_4 or $(NH_2)_2CO$. Urea is essentially a waste product: it has no physiological function. It is dissolved in blood and excreted by the kidney.
Euphotic zone	The surface layer of water to the depth of light penetration at which photosynthesis balances respiration is an euphotic zone.
Decomposition	Decomposition refers to the reduction of the body of a formerly living organism into simpler forms of matter.
Uric acid	An insoluble precipitate of nitrogenous waste excreted by land snails, insects, birds, and some reptiles is called uric acid.

Nutrient regeneration	Nutrient regeneration refers to the release of nutrients from organic matter by decomposers.
Habitat	Habitat refers to a place where an organism lives; an environmental situation in which an organism lives.
Isotope	An isotope is a form of an element whose nuclei have the same atomic number - the number of protons in the nucleus - but different mass numbers because they contain different numbers of neutrons.
Ammonia	Ammonia is a compound of nitrogen and hydrogen with the formula NH_3. At standard temperature and pressure ammonia is a gas. It is toxic and corrosive to some materials, and has a characteristic pungent odor.
Gyre	Gyre refers to a roughly circular pattern of ocean currents, formed because continents interrupt the currents' flow; rotates clockwise in the Northern Hemisphere and counterclockwise in the Southern Hemisphere.
Atmosphere	Earth's atmosphere is a layer of gases surrounding the planet Earth and retained by the Earth's gravity. It contains roughly 78% nitrogen and 21% oxygen, with trace amounts of other gases.
Nitrifying bacteria	Several kinds of bacteria capable of converting ammonia to nitrite, or nitrite to nitrate is referred to as nitrifying bacteria.
Ion	Ion refers to an atom or molecule that has gained or lost one or more electrons, thus acquiring an electrical charge.
Denitrifying bacteria	Denitrifying bacteria form a necessary part of the process known as denitrification as part of the nitrogen cycle, their primary purpose being to metabolise nitrogenous compounds, with the assistance of the nitrase reductase enzyme, to turn oxides back to nitrogen gas or nitrous oxides for energy generation.
Anaerobic	An anaerobic organism is any organism that does not require oxygen for growth.
Nitrogen fixation	Nitrogen fixation is the process by which nitrogen is taken from its relatively inert molecular form (N_2) in the atmosphere and converted into nitrogen compounds useful for other chemical processes.
Cyanobacteria	Cyanobacteria are a phylum of bacteria that obtain their energy through photosynthesis. They are often referred to as blue-green algae, even though it is now known that they are not directly related to any of the other algal groups, which are all eukaryotes.
Food chain	Food chain refers to a sequence of food transfers from one species to another within an ecosystem.
Yeast	Yeast refers to common term for several families of unicellular fungi. Includes species used for brewing beer and making bread, as well as pathogenic species.
Denitrification	Conversion of nitrate or nitrite to nitrogen gas by bacteria in soil is denitrification.
Phosphate	A phosphate is a polyatomic ion or radical consisting of one phosphorus atom and four oxygen. In the ionic form, it carries a -3 formal charge, and is denoted PO_4^{3-}.
Adenosine triphosphate	Adenosine triphosphate is the nucleotide known in biochemistry as the "molecular currency" of intracellular energy transfer; that is, able to store and transport chemical energy within cells.
Grazing	Grazing refers to consumption of a portion of a plant's tissues, generallyapplied to grasses and other herbaceous vegetation; removing plant or algal cover from a substrate, generally applied to aquatic systems.

Sediment	Sediment is any particulate matter that can be transported by fluid flow and which eventually is deposited as a layer of solid particles on the bed or bottom of a body of water or other liquid.
Diffusion	Diffusion refers to the spontaneous movement of particles of any kind from where they are more concentrated to where they are less concentrated.
Growth rate	A measure of the change in population size per individual per unit of time is called growth rate.
Crop	An organ, found in both earthworms and birds, in which ingested food is temporarily stored before being passed to the gizzard, where it is pulverized is the crop.
Skeleton	In biology, the skeleton or skeletal system is the biological system providing physical support in living organisms.
Silicon	Silicon is the chemical element in the periodic table that has the symbol Si and atomic number 14. It is the second most abundant element in the Earth's crust, making up 25.7% of it by weight.
Cell division	Cell division is the process by which a cell (called the parent cell) divides into two cells (called daughter cells). Cell division is usually a small segment of a larger cell cycle. In meiosis, however, a cell is permanently transformed and cannot divide again.
Molybdenum	Molybdenum is a chemical element in the periodic table. Its symbol is Mo and its atomic number 42. Molybdenum is a transition metal. The pure metal is silvery white in color and very hard, and has one of the highest melting points of all pure elements.
Manganese	Manganese is a chemical element in the periodic table that has the symbol Mn and atomic number 25.
Cofactor	A cofactor is any substance that needs to be present in addition to an enzyme to catalyze a certain reaction.
Cobalt	Cobalt is a chemical element in the periodic table that has the symbol Co and atomic number 27. Cobalt is a hard ferromagnetic silver-white element.
Zinc	Zinc is a chemical element in the periodic table that has the symbol Zn and atomic number 30.
Pleistocene	A geologic epoch, characterized by alternating glacial and interglacial stages, that ended about 10,000 years ago, that lasted for 2 million years is called the pleistocene epoch.
Desert	A desert is a landscape form or region that receives little precipitation - less than 250 mm (10 in) per year. It is a biome characterized by organisms adapted to sparse rainfall and rapid evaporation.
Vitamin	A Vitamin is an organic molecule required by a living organism in minute amounts for proper health. An organism deprived of all sources of a particular vitamin will eventually suffer from disease symptoms specific to that vitamin.
Thiamine	Thiamine, also known as vitamin B_1, is a colorless compound with chemical formula $C_{12}H_{17}ClN_4OS$. Systemic thiamine deficiency can lead to myriad problems including neurodegeneration, wasting, and death. Well-known syndromes caused by lack of thiamine due to malnutrition or a diet high in thiaminase-rich foods include Wernicke-Korsakoff syndrome and beriberi, diseases also common in chronic abusers of alcohol.
Biotin	Biotin is a low-molecular-weight compound used as a coenzyme. Useful technically as a covalent label for proteins, allowing them to be detected by the egg protein avidin, which binds extremely tightly to biotin.
Coccolithophore	A coccolithophore is a single-celled algae, or phytoplankton, belonging to the haptophytes.

	It is distinguished by special calcium carbonate plates (or scales) of unknown purpose called coccoliths.
Sargasso sea	Sargasso sea refers to the region of the Atlantic Ocean north of the West Indies that is characterized by floating masses of Sargasso weed, a brown alga.
Dominance	Dominance is the state of having high social status relative to other individuals, who react submissively to dominant individuals.
Exudate	An exudate is any fluid that filters from the circulatory system into leisions or areas of inflamation. Its composition varies but generally includes water and the disolved solutes of the blood, some or all plasma proteins, white blood cells, platelets and red blood cells.
Silica	Silica is a mineral similar to glass that is the major component of the cell wall, shell, or skeleton of many marine organisms.
Allelopathy	The term allelopathy denotes the production of specific biomolecules by one plant that can induce suffering in, or give benefit to, another plant.
Relative abundance	Differences in the abundance of species within a community is relative abundance.
Cyst	A cyst is a closed sac having a distinct membrane and developing abnormally in a cavity or structure of the body. They may occur as a result of a developmental error in the embryo during pregnancy or they may be caused by infections.
Spine	A spine is a rigid, pointed surface protuberance or needle-like structure on an animal, shell, or plant, presumably serving as a defense against attack by predators.
Boundary layer	Boundary layer refers to a layer of still or slow-moving water or air close to the surface of an object.
Oligotrophic	Oligotrophic refers to any environment which offers little to sustain life. This term is usually used to describe bodies of water with very low nutrient levels.
Microbial loop	Microbial loop refers to the mineralization of organic matter synthesized by photosynthetic phytoplankton through the activity of microorganisms such as bacteria and protozoa.
Lactic acid	Lactic acid accumulates in skeletal muscles during extensive anaerobic exercise, causing temporary muscle pain. Lactic acid is quickly removed from muscles when they resume aerobic metabolism.
Alcohol	Alcohol is a general term, applied to any organic compound in which a hydroxyl group (-OH) is bound to a carbon atom, which in turn is bound to other hydrogen and/or carbon atoms. The general formula for a simple acyclic alcohol is $C_nH_{2n+1}OH$.
Particulate matter	In air pollution, solid particles and liquid droplets, as opposed to material uniformly dispersed among air molecules is particulate matter.
Phagocytosis	Phagocytosis is a form of endocytosis where large particles are enveloped by the cell membrane of a (usually larger) cell and internalized to form a phagosome, or "food vacuole."
Complement	Complement is a group of proteins of the complement system, found in blood serum which act in concert with antibodies to achieve the destruction of non-self particles such as foreign blood cells or bacteria.
Cellulose	A large polysaccharide composed of many glucose monomers linked into cable-like fibrils that provide structural support in plant cell walls is referred to as cellulose.
Chitin	Chitin refers to strong but flexible nitrogenous polysaccharide found in fungal cells and in the exoskeleton of arthropods.

Herbivore A herbivore is an animal that is adapted to eat primarily plant matter

Virus Obligate intracellular parasite of living cells consisting of an outer capsid and an inner core of nucleic acid is referred to as virus. The term virus usually refers to those particles that infect eukaryotes whilst the term bacteriophage or phage is used to describe those infecting prokaryotes.

Algae The algae consist of several different groups of living organisms that capture light energy through photosynthesis, converting inorganic substances into simple sugars with the captured energy.

Substrate A substrate is a molecule which is acted upon by an enzyme. Each enzyme recognizes only the specific substrate of the reaction it catalyzes. A surface in or on which an organism lives.

Radiolarian Radiolarian are amoeboid protozoa that produce intricate mineral skeletons, typically with a central capsule dividing the cell into inner and outer portions, called endoplasm and ectoplasm. They are found as plankton throughout the ocean, and their shells are important fossils found from the Cambrian onwards.

Ciliate Ciliate is one of the most important groups of protists, common almost everywhere there is water. They are a type of protozoan that moves by means of cilia.

Food web A network of interconnecting food chains is referred to as a food web.

Trophic level In ecology, the trophic level is the position that an organism occupies in a food chain - what it eats, and what eats it.

Atom An atom is the smallest possible particle of a chemical element that retains its chemical properties.

Prediction Step of the scientific method which follows the formulation of a hypothesis and assists in creating the experimental design is referred to as prediction.

Clone A group of genetically identical cells or organisms derived by asexual reproduction from a single parent is called a clone.

Coexistence Coexistence refers to occurrence of two or more species in the same habitat; usually applied to potentially competing species.

Competitive exclusion Competitive exclusion refers to the elimination of one species by another as a result of competition.

Blood Blood is a circulating tissue composed of fluid plasma and cells. The main function of blood is to supply nutrients (oxygen, glucose) and constitutional elements to tissues and to remove waste products.

Algal bloom An algal bloom is a relatively rapid increase in the population of phytoplankton algae in an aquatic system. Typically only one or a few species are involved and the bloom is recognized by discoloration of the water resulting from the high density of pigmented cells.

Red tide Red tide is estuarine or marine algal bloom. Red tide is caused by species of dinoflagellates, often present in sufficient numbers (thousands or millions of cells per milliliter) to turn the water red or brown.

Toxin Toxin refers to a microbial product or component that can injure another cell or organism at low concentrations. Often the term refers to a poisonous protein, but toxins may be lipids and other substances.

Metabolic rate Energy expended by the body per unit time is called metabolic rate.

Mollusk The mollusk is the large and diverse phylum Mollusca, which includes a variety of familiar creatures well-known for their decorative shells or as seafood. These range from tiny snails,

	clams, and abalone to the octopus, cuttlefish and squid (which are considered the most intelligent invertebrates).
Dormancy	Dormancy is a period of arrested plant growth. It is a survival strategy exhibited by many plant species, which enables them to survive in climates where part of the year is unsuitable for growth.
Generation time	Generation time refers to the time between the birth of a parent and the birth of its offspring. The time interval between successive cell divisions.
Reproduction	Biological reproduction is the biological process by which new individual organisms are produced. Reproduction is a fundamental feature of all known life; each individual organism exists as the result of reproduction by an antecedent.
Standing stock	Standing stock refers to the total amount, or biomass, of an organism at a given time.
Digestion	Digestion refers to the mechanical and chemical breakdown of food into molecules small enough for the body to absorb; the second main stage of food processing, following ingestion.
Threshold	Electrical potential level at which an action potential or nerve impulse is produced is called threshold.
Tintinnids	Ciliates that secrete vase-like cases, or loricas are tintinnids.
Biogeochemical cycle	A biogeochemical cycle is a circuit or pathway by which a chemical element or molecule moves through both biotic ("bio-") and abiotic ("geo-") compartments of an ecosystem.
Ecosystem	In general terms an ecosystem can be thought of as an assemblage of organisms (plant, animal and other living organisms living together with their environment, functioning as a loose unit. That is, a dynamic and complex whole, interacting as an "ecological unit".
Ecology	Ecology is the scientific study of the distribution and abundance of living organisms and how these properties are affected by interactions between the organisms and their environment.
Filter feeder	A Filter feeder is an animal that feeds by straining suspended matter and food particles from water, typically by passing the water over a specialized structure, such as the baleen of baleen whales.
Biology	Biology is the branch of science dealing with the study of life. It is concerned with the characteristics, classification, and behaviors of organisms, how species come into existence, and the interactions they have with each other and with the environment.
Iron deficiency	Iron deficiency (or "sideropenia") is the most common known form of nutritional deficiency.
Microbe	A microbe is an organism that is so small that it is microscopic (invisible to the naked eye). Microorganisms are often illustrated using single-celled, or unicellular organisms; however, some unicellular protists are visible to the naked eye, and some multicellular species are microscopic.
Systematics	In biology, systematics is the study of the diversity of organism characteristics, and especially how they relate evolutionarily. The term "systematics" may or may not overlap with "taxonomy" which concerns itself with scientific classification of species and other taxa.
Bacterioplankton	Bacterioplankton refers to the bacterial component of the plankton that drifts in the water column. Most obtain their energy through decomposition of other organisms, which renders them dependent on the phytoplankton for the production of dissolved organic matter as their main food source.
Sulfide	One of the minerals that is abundant in the hot water that seeps through hydrothermal vents is sulfide.
Predation	Interaction in which one organism uses another, called the prey, as a food source is referred

to as predation.

Protozoa

Protozoa are single-celled eukaryotes (organisms with nuclei) that show some characteristics usually associated with animals, most notably mobility and heterotrophy. They are often grouped in the kingdom Protista together with the plant-like algae.

Limnology

Limnology is a discipline that concerns the study of inland waters (both saline and fresh), specifically lakes, ponds and rivers (both natural and manmade), including their biological, physical, chemical, and hydrological aspects.

Genus

In biology, a genus is a taxonomic grouping. That is, in the classification of living organisms, a genus is considered to be distinct from other such genera. A genus has one or more species: if it has more than one species these are likely to be morphologically more similar than species belonging to different genera.

Water column	The vertical column of seawater that extends from the surface to the bottom is called water column.
Coral reef	A coral reef is a type of biotic reef that develops in tropical waters by the growth of coralline algae, hermatypic corals, and other marine organisms.
Food web	A network of interconnecting food chains is referred to as a food web.
Tissue	Group of similar cells which perform a common function is called tissue.
Carbon	Carbon is a chemical element in the periodic table that has the symbol C and atomic number 6. An abundant nonmetallic, tetravalent element, carbon has several allotropic forms.
Primary productivity	Primary productivity is the amount of light energy converted to chemical energy in a given amount of time. It usually corresponds to the amount of photosynthesis that occurred.
Photosynthesis	Photosynthesis is a biochemical process in which plants, algae, and some bacteria harness the energy of light to produce food. Ultimately, nearly all living things depend on energy produced from photosynthesis for their nourishment, making it vital to life on Earth.
Biomass	Biomass is organic non-fossil material, collectively. In other words, biomass comprises the mass of all biological organisms, dead or alive, excluding biological mass that has been transformed by geological processes into substances such as coal or petroleum.
Secondary production	Secondary production is the biomass production resulting from the assimilation of organic matter produced by a primary consumer; production by organisms (mainly animals) which consume primary producers (mainly plants).
Phytoplankton	Phytoplankton refers to the autotrophic component of the plankton that drifts in the water column. Phytoplankton obtain energy through a process called photosynthesis, and so must live in the well-lit surface layer (termed the euphotic zone) of an ocean, sea, or lake.
Cell	The cell is the structural and functional unit of all living organisms, and is sometimes called the "building block of life."
Fatty acid	A fatty acid is a carboxylic acid (or organic acid), often with a long aliphatic tail (long chains), either saturated or unsaturated.
Vitamin	A Vitamin is an organic molecule required by a living organism in minute amounts for proper health. An organism deprived of all sources of a particular vitamin will eventually suffer from disease symptoms specific to that vitamin.
Heterotrophic	Incapable of synthesizing food and therefore dependent on other organisms for it is called heterotrophic.
Bacteria	The domain that contains procaryotic cells with primarily diacyl glycerol diesters in their membranes and with bacterial rRNA. Bacteria also is a general term for organisms that are composed of procaryotic cells and are not multicellular.
Food chain	Food chain refers to a sequence of food transfers from one species to another within an ecosystem.
Flagellate	A flagellate is a cell with one or more whip-like organelles called flagella. Some cells in animals may be flagellate, for instance the sperm cells of most phyla. In protists and microscopic animals, flagella are generally used for propulsion.
Tunicate	One of a group of invertebrate chordates in the subphylum of saclike filter feeders with input and output siphons is referred to as a tunicate.
Barnacle	A barnacle is a type of arthropod belonging to infraclass Cirripedia in the subphylum Crustacea and is hence distantly related to crabs and lobsters. They live attached to surfaces and are typically enclosed by heavy calcareous plates are referred to as barnacles.

Copepods	A group of small crustaceans that are important members of marine and freshwater plankton communities is copepods.
Copepod	The group of small crustaceans that is an important member of marine and freshwater plankton communities is copepod. They are very important food organisms for small fish, whales, and other crustaceans such as krill in the ocean and in fresh water.
Diatom	Diatom refers to a unicellular photosynthetic alga with a unique, glassy cell wall containing silica.
Species	Group of similarly constructed organisms capable of interbreeding and producing fertile offspring is a species.
Trophic level	In ecology, the trophic level is the position that an organism occupies in a food chain - what it eats, and what eats it.
Crustacean	Member of a group of marine arthropods that contains among others shrimps, crabs, crayfish, and lobsters is called crustacean.
Predator	A predator is an animal or other organism that hunts and kills other organisms for food in an act called predation.
Zooplankton	Nonphotosynthetic protists that feed on other plankton and are abundant in marine and freshwater environments are zooplankton.
Egg	An egg is the zygote, resulting from fertilization of the ovum. It nourishes and protects the embryo.
Spine	A spine is a rigid, pointed surface protuberance or needle-like structure on an animal, shell, or plant, presumably serving as a defense against attack by predators.
Toxin	Toxin refers to a microbial product or component that can injure another cell or organism at low concentrations. Often the term refers to a poisonous protein, but toxins may be lipids and other substances.
Respiration	Respiration is the process by which an organism obtains energy by reacting oxygen with glucose to give water, carbon dioxide and ATP (energy). Respiration takes place on a cellular level in the mitochondria of the cells and provide the cells with energy.
Primary production	Primary production is the production of organic compounds from inorganic materials principally through the process of photosynthesis.
Upwelling	Upward movement of deep, nutrient-rich water along coasts is referred to as upwelling.
Ecosystem	In general terms an ecosystem can be thought of as an assemblage of organisms (plant, animal and other living organisms living together with their environment, functioning as a loose unit. That is, a dynamic and complex whole, interacting as an "ecological unit".
Prediction	Step of the scientific method which follows the formulation of a hypothesis and assists in creating the experimental design is referred to as prediction.
Gross primary productivity	Gross primary productivity refers to production before respiration losses are subtracted; photosynthetic production for plants and metabolizable production for animals.
Oxygen	Oxygen is a chemical element in the periodic table. It has the symbol O and atomic number 8. Oxygen is the second most common element on Earth, composing around 46% of the mass of Earth's crust and 28% of the mass of Earth as a whole, and is the third most common element in the universe.
Plankton	Plankton are drifting organisms that inhabit the water column of oceans, seas, and bodies of fresh water. Plankton abundance and distribution are strongly dependent on factors such as ambient nutrients concentrations, the physical state of the water column, and the abundance

	of other plankton.
Dissolved oxygen	Oxygen gas molecules dissolved in water. Fish and other aquatic organisms are dependent on dissolved oxygen for respiration. Therefore concentration of dissolved oxygen is a measure of water quality.
Electrode	An electrode is a conductor used to make contact with a nonmetallic part of a circuit (e.g. a semiconductor, an electrolyte or a vacuum).
Radioactive	A term used to describe the property of releasing energy or particles from an unstable atom is called radioactive.
Isotype	Isotype refers to a variant form of an immunoglobulin that occurs in every normal individual of a particular species. Usually the characteristic antigenic determinant is in the constant region of H and L chains.
Nitrogen	A colorless and tasteless and mostly inert diatomic non-metal gas that is an essential constituent of proteins is nitrogen.
Ammonia	Ammonia is a compound of nitrogen and hydrogen with the formula NH_3. At standard temperature and pressure ammonia is a gas. It is toxic and corrosive to some materials, and has a characteristic pungent odor.
Protein	A protein is a complex, high-molecular-weight organic compound that consists of amino acids joined by peptide bonds. They are essential to the structure and function of all living cells and viruses. Many are enzymes or subunits of enzymes.
Nitrate	Nitrate refers to a salt of nitric acid; a compound containing the radical NO_3; biologically, the final form of nitrogen from the oxidation of organic nitrogen compounds.
Hexose	A hexose is a monosaccharide with six carbon atoms having the chemical formula $C_6H_{12}O_6$.
Lipid	Lipid is one class of aliphatic hydrocarbon-containing organic compounds essential for the structure and function of living cells. They are characterized by being water-insoluble but soluble in nonpolar organic solvents.
Sugar	A sugar is the simplest molecule that can be identified as a carbohydrate. These include monosaccharides and disaccharides, trisaccharides and the oligosaccharides. The term "glyco-" indicates the presence of a sugar in an otherwise non-carbohydrate substance.
Particulate matter	In air pollution, solid particles and liquid droplets, as opposed to material uniformly dispersed among air molecules is particulate matter.
Standing stock	Standing stock refers to the total amount, or biomass, of an organism at a given time.
Net primary production	Net primary production is the energy or biomass content of plant material that has accumulated in an ecosystem over a period of time through photosynthesis. It is the amount of energy left after subtracting the respiration of primary producers (mostly plants) from the total amount of energy (mostly solar) that is fixed biologically.
Photosynthetic pigment	A photosynthetic pigment is a pigment that is present in chloroplasts or photosynthetic bacteria and captures the light energy necessary for photosynthesis.
Surface water	Includes all bodies of water, lakes, rivers, ponds, and so on that are on the surface of the earth, in contrast to groundwater, which lies below the surface is surface water.
Chlorophyll	Chlorophyll is a green photosynthetic pigment found in plants, algae, and cyanobacteria. In plant photosynthesis incoming light is absorbed by chlorophyll and other accessory pigments in the antenna complexes of photosystem I and photosystem II.
Pigment	Pigment is any material resulting in color in plant or animal cells which is the result of selective absorption.

Fluorescence	Fluorescence is a luminescence that is mostly found as an optical phenomenon in cold bodies, in which the molecular absorption of a photon triggers the emission of a lower-energy photon with a longer wavelength. The energy difference between the absorbed and emitted photons ends up as molecular vibrations or heat.
Electron	The electron is a light fundamental subatomic particle that carries a negative electric charge. The electron is a spin-1/2 lepton, does not participate in strong interactions and has no substructure.
Distribution	Distribution in pharmacology is a branch of pharmacokinetics describing reversible transfer of drug from one location to another within the body.
Coriolis effect	The deflection of large bodies of water or air to the right in the Northern Hemisphere and to the left in the Southern Hemisphere is referred to as the coriolis effect.
Trade winds	The trade winds are a pattern of wind found in bands around the Earth's equatorial region. The trade winds are the prevailing winds in the tropics, blowing from the high-pressure area in the horse latitudes towards the low-pressure area around the equator.
Crop	An organ, found in both earthworms and birds, in which ingested food is temporarily stored before being passed to the gizzard, where it is pulverized is the crop.
Population	Group of organisms of the same species occupying a certain area and sharing a common gene pool is referred to as population.
Guano	Guano is the name given to the collected droppings of seabirds and bats. It is highly prized as an effective fertilizer due to its high levels of phosphorus and nitrogen.
Monsoon	A monsoon is a wind pattern that reverses direction on a seasonal basis. The term was originally applied to monsoonal winds in the Arabian Sea and Indian Ocean.
Benthos	Benthos are the organisms and habitats of the sea floor; in freshwater biology they are the organisms and habitats of the bottoms of lakes, rivers, and creeks.
Seaweed	Phycologists consider seaweed to refer any of a large number of marine benthic algae that are multicellular, macrothallic (large-bodied), and thus differentiated from most algae that tend to be microscopic in size.
Zostera	Zostera is a small genus of widely distributed seagrass, commonly called eelgrass. It contains twelve species. Zostera is found on sandy sunstrates or in esturaries submerged or partially floating. Most Zostera are perennial, they have long, bright green, ribbon-like leaves, about 1 cm wide.
Continental shelf	The shallow part of the seafloor immediately adjacent to a continent is called the continental shelf.
Surface layer	The upper layer of water that is mixed by wind, waves, and currents is the surface layer .
Invertebrate	Invertebrate is a term coined by Jean-Baptiste Lamarck to describe any animal without a spinal column. It therefore includes all animals except vertebrates (fish, reptiles, amphibians, birds and mammals).
Mammal	Homeothermic vertebrate characterized especially by the presence of hair and mammary glands is a mammal.
Barren	An area with sparse vegetation owing to some physical or chemical property of the soil is referred to as barren.
Gyre	Gyre refers to a roughly circular pattern of ocean currents, formed because continents interrupt the currents' flow; rotates clockwise in the Northern Hemisphere and counterclockwise in the Southern Hemisphere.

Sargasso sea	Sargasso sea refers to the region of the Atlantic Ocean north of the West Indies that is characterized by floating masses of Sargasso weed, a brown alga.
Growing season	Growing season refers to the period of the year during which conditions are suitable for plant growth; in temperate regions, generally between the first and last frosts.
Thermocline	The thermocline is a layer within a body of water where the temperature changes rapidly with depth. Because water is not perfectly transparent, almost all sunlight is absorbed in the surface layer, which heats up.
Stratification	Stratification is the process of pretreating seeds to simulate natural conditions that a seed must endure before germination. Many seed species have what is called an embryonic dormancy and generally speaking will not sprout until this dormancy is broken.
Predation	Interaction in which one organism uses another, called the prey, as a food source is referred to as predation.
Iron	Iron is essential to all organisms, except for a few bacteria. It is mostly stably incorporated in the inside of metalloproteins, because in exposed or in free form it causes production of free radicals that are generally toxic to cells.
Oxygen minimum layer	Oxygen minimum layer refers to a layer of water at a depth of approximately 500 m where the concentration of dissolved oxygen is very low.
Net primary productivity	The energy stored in the autotrophs of an ecosystem over a given time period is referred to as net primary productivity.
Phosphorus	Phosphorus is the chemical element in the periodic table that has the symbol P and atomic number 15.
Euphotic zone	The surface layer of water to the depth of light penetration at which photosynthesis balances respiration is an euphotic zone.
Ecology	Ecology is the scientific study of the distribution and abundance of living organisms and how these properties are affected by interactions between the organisms and their environment.
Biogeochemical cycle	A biogeochemical cycle is a circuit or pathway by which a chemical element or molecule moves through both biotic ("bio-") and abiotic ("geo-") compartments of an ecosystem.
Biosphere	The biosphere is that part of a planet's outer shell — including air, land, surface rocks and water — within which life occurs, and which biotic processes in turn alter or transform.
Fertilization	Fertilization is fusion of gametes to form a new organism. In animals, the process involves a sperm fusing with an ovum, which eventually leads to the development of an embryo.
Salt marsh	Salt marsh refers to a grassy area that extends along the shores of estuaries and sheltered coasts in temperate and subpolar regions.

Invertebrate	Invertebrate is a term coined by Jean-Baptiste Lamarck to describe any animal without a spinal column. It therefore includes all animals except vertebrates (fish, reptiles, amphibians, birds and mammals).
Evolution	In biology, evolution is the process by which novel traits arise in populations and are passed on from generation to generation. Its action over large stretches of time explains the origin of new species and ultimately the vast diversity of the biological world.
Cell	The cell is the structural and functional unit of all living organisms, and is sometimes called the "building block of life."
Tissue	Group of similar cells which perform a common function is called tissue.
Large intestine	In anatomy of the digestive system, the colon, also called the large intestine or large bowel, is the part of the intestine from the cecum ('caecum' in British English) to the rectum. Its primary purpose is to extract water from feces.
Liver	The liver is an organ in vertebrates, including humans. It plays a major role in metabolism and has a number of functions in the body including drug detoxification, glycogen storage, and plasma protein synthesis. It also produces bile, which is important for digestion.
Organ	Organ refers to a structure consisting of several tissues adapted as a group to perform specific functions.
Nerve	A nerve is an enclosed, cable-like bundle of nerve fibers or axons, which includes the glia that ensheath the axons in myelin.
Skin	Skin is an organ of the integumentary system composed of a layer of tissues that protect underlying muscles and organs.
Transformation	Transformation is the genetic alteration of a cell resulting from the introduction, uptake and expression of foreign genetic material (DNA or RNA).
Symmetry	Symmetry in biology is the balanced distribution of duplicate body parts or shapes.
Sea anemone	Named after a terrestrial flower, the sea anemone is a group of water-dwelling, filter feeding animals of the order Actinaria. As a cnidarian, it is closely related to coral and jellyfish.
Body cavity	A fluid-containing space between the digestive tract and the body wall is referred to as body cavity.
Coelom	A fluid filled body cavity with a complete lining called peritoneum derived from mesoderm is called a coelom. Organs formed inside a coelom can freely move, grow, and develop independently of the body wall while fluid cushions and protects them from shocks.
Pseudocoel	Body cavity containing a fluid in which the various internal organs are suspended is a pseudocoel.
Mesoderm	Mesoderm forms in the embryos of animals more complex than cnidarians. Some of the cells migrating inward to form the endoderm form an additional layer between the other two. It gives rise to muscles, bones, the dermis of the skin, and most other organs in the adult.
Element	A chemical element, often called simply element, is a chemical substance that cannot be divided or changed into other chemical substances by any ordinary chemical technique. An element is a class of substances that contain the same number of protons in all its atoms.
Species	Group of similarly constructed organisms capable of interbreeding and producing fertile offspring is a species.
Phylum	Phylum is the second highest taxonomic classification for the kingdom Animalia (animals), between kingdom level and class level.

Morphology	The scientific study of organic form, including both its development and function is morphology.
Zoology	Zoology is the biological discipline which involves the study of animals.
Circulatory system	The circulatory system or cardiovascular system is the organ system which circulates blood around the body of most animals.
Nervous system	The nervous system of an animal coordinates the activity of the muscles, monitors the organs, constructs and processes input from the senses, and initiates actions.
Segmentation	Segmentation in biology refers to the division of some metazoan bodies and plant body plans into a series of semi-repetitive segments, and the question of the benefits and costs of doing so.
Excretion	Excretion is the biological process by which an organism chemically separates waste products from its body. The waste products are then usually expelled from the body by elimination.
Diffusion	Diffusion refers to the spontaneous movement of particles of any kind from where they are more concentrated to where they are less concentrated.
Protista	A taxonomic kingdom including unicellular, eukaryotic organisms is a protista. A heterogeneous group of living things, comprising those eukaryotes that are neither animals, plants, nor fungi.
Protist	A protist is a heterogeneous group of living things, comprising those eukaryotes that are neither animals, plants, nor fungi. They are a paraphyletic grade, rather than a natural group, and do not have much in common besides a relatively simple organization
Ciliate	Ciliate is one of the most important groups of protists, common almost everywhere there is water. They are a type of protozoan that moves by means of cilia.
Amoeba	Amoeba is a genus of protozoa that moves by means of temporary projections called pseudopods, and is well-known as a representative unicellular organism. Amoeba itself is found in freshwater, typically on decaying vegetation from streams, but is not especially common in nature.
Host	Host is an organism that harbors a parasite, mutual partner, or commensal partner; or a cell infected by a virus.
Cytoplasm	Cytoplasm refers to everything inside a cell between the plasma membrane and the nucleus; consists of a semifluid medium and organelles.
Organelle	In cell biology, an organelle is one of several structures with specialized functions, suspended in the cytoplasm of a eukaryotic cell.
Nucleus	In cell biology, the nucleus is found in all eukaryotic cells that contains most of the cell's genetic material. The nucleus has two primary functions: to control chemical reactions within the cytoplasm and to store information needed for cellular division.
Bacteria	The domain that contains procaryotic cells with primarily diacyl glycerol diesters in their membranes and with bacterial rRNA. Bacteria also is a general term for organisms that are composed of procaryotic cells and are not multicellular.
Amoeboid	Amoeboid is a cell that moves or feeds by means of temporary projections, called pseudopods. They have appeared in a number of different groups. Some cells in multicellular animals may be amoeboid, for instance our white blood cells, which consume pathogens
Cilia	Numerous short, hairlike structures projecting from the cell surface that enable locomotion are cilia.
Calcium	A mineral that is the major component of the shell, skeleton, and other parts of many

carbonate	organisms are called calcium carbonate.
Foraminifera	The Foraminifera, or forams for short, are a large group of amoeboid protists with reticulating pseudopods, fine strands that branch and merge to form a dynamic net. They typically produce a shell, or test, which can have either one or multiple chambers, some becoming quite elaborate in structure.
Skeleton	In biology, the skeleton or skeletal system is the biological system providing physical support in living organisms.
Silica	Silica is a mineral similar to glass that is the major component of the cell wall, shell, or skeleton of many marine organisms.
Protoplasm	Protoplasm is the living substance inside the cell. At the simplest level, it is divisible into cytoplasm and nucleoplasm.
Porifera	The sponges or poriferans are animals of the phylum Porifera. They are primitive, sessile, mostly marine, waterdwelling filter feeders that pump water through their bodies to filter out particles of food matter.
Sponge	An invertebrates that consist of a complex aggregation of cells, including collar cells, and has a skeleton of fibers and/or spicules is a sponge. They are primitive, sessile, mostly marine, waterdwelling filter feeders that pump water through their matrix to filter out particulates of food matter.
Deep sea	The deep sea refers to those areas of oceans to which little or no light penetrates.
Predator	A predator is an animal or other organism that hunts and kills other organisms for food in an act called predation.
Spine	A spine is a rigid, pointed surface protuberance or needle-like structure on an animal, shell, or plant, presumably serving as a defense against attack by predators.
Zooxanthellae	Zooxanthellae refers to dinoflagellates and other algae that live as symbionts within coral.
Algae	The algae consist of several different groups of living organisms that capture light energy through photosynthesis, converting inorganic substances into simple sugars with the captured energy.
Sugar	A sugar is the simplest molecule that can be identified as a carbohydrate. These include monosaccharides and disaccharides, trisaccharides and the oligosaccharides. The term "glyco-" indicates the presence of a sugar in an otherwise non-carbohydrate substance.
Radiolarian	Radiolarian are amoeboid protozoa that produce intricate mineral skeletons, typically with a central capsule dividing the cell into inner and outer portions, called endoplasm and ectoplasm. They are found as plankton throughout the ocean, and their shells are important fossils found from the Cambrian onwards.
Genus	In biology, a genus is a taxonomic grouping. That is, in the classification of living organisms, a genus is considered to be distinct from other such genera. A genus has one or more species: if it has more than one species these are likely to be morphologically more similar than species belonging to different genera.
Sediment	Sediment is any particulate matter that can be transported by fluid flow and which eventually is deposited as a layer of solid particles on the bed or bottom of a body of water or other liquid.
Habitat	Habitat refers to a place where an organism lives; an environmental situation in which an organism lives.
Flagellate	A flagellate is a cell with one or more whip-like organelles called flagella. Some cells in

	animals may be flagellate, for instance the sperm cells of most phyla. In protists and microscopic animals, flagella are generally used for propulsion.
Collar cell	Collar cell refers to a specialized cell lining the inside channels of sponges. Flagella extend from a sievelike collar, creating a water current that draws microscopic organisms through the collar and into the body, where they become trapped.
Choanocyte	Choanocyte refers to a flagellated feeding cell found in sponges. Also called a collar cell, it has a collarlike ring that traps food particles around the base of its flagellum.
Epithelial	Functions of epithelial cells include secretion, absorption, protection, transcellular transport, sensation detection, and selective permeability.
Colony	Colony refers to a cluster or assemblage of microorganisms growing on the surface of an agar culture medium. A colony also refers to several individual organisms of the same species living closely together, usually for mutual benefit, such as stronger defences, the ability to attack bigger prey, etc.
Amoebocyte	Amoebocyte refers to an amoebalike cell that moves by pseudopodia. Depending on the species, it may digest and distribute food, dispose of wastes, form skeletal fibers, fight infections, and change into other cell types.
Osculum	Osculum is a large opening to the outside, in which the current of water exits after passing through the large cavity (spongocoel) of a sponge.
Calcareous	Calcareous refers to a sediment, sedimentary rock, or soil type which is formed from or contains a high proportion of calcium carbonate in the form of calcite or aragonite. It is also used to refer to relatively alkaline soil.
Spicule	Spicule refers to a subunit of the endoskeleton of sponges that is made of silica or calcium carbonate.
Asexual reproduction	Asexual reproduction refers to reproduction that does not involve the fusion of haploid sex cells. The parent body may divide and new parts regenerate, or a new, smaller individual may form as an attachment to the parent, to drop off when complete.
Versatile	The anther can be attached to the filament in two ways, versatile is when it is attached at its center to the filament; pollen is then released through pores (poricidal dehiscence).
Sperm	Sperm refers to the male sex cell with three distinct parts at maturity: head, middle piece, and tail.
Population	Group of organisms of the same species occupying a certain area and sharing a common gene pool is referred to as population.
Coral reef	A coral reef is a type of biotic reef that develops in tropical waters by the growth of coralline algae, hermatypic corals, and other marine organisms.
Nerve cell	A cell specialized to originate or transmit nerve impulses is referred to as nerve cell.
Hydrozoans	Colonial hydrozoans have both a medusa stage and a polyp stage in their life cycle. They have a base, a stalk, and one or more polyps. Most polyps are feeding polyps called hydranths. They prey on tiny crustaceans, worms, and larvae.
Jellyfish	Jellyfish are marine invertebrates belonging to the Scyphozoa class, and in turn the phylum Cnidaria. The body of an adult jellyfish is composed of a bell-shaped, jellylike substance enclosing its internal structure, from which the creature's tentacles suspend.
Cnidaria	Cnidaria is a phylum containing some 10,000 species of relatively simple animals found exclusively in aquatic environments (most species are marine). They get their name from cnidocytes, which are specialized cells that carry stinging organelles.

Coral	Group of cnidarians having a calcium carbonate skeleton that participate in the formation of reefs is coral.
Digestive tract	The digestive tract is the system of organs within multicellular animals which takes in food, digests it to extract energy and nutrients, and expels the remaining waste.
Tentacle	An elongate, extensible projection of the body of cnidarians and cephalopod mollusks that may be used for grasping, stinging, and immobilizing prey, and for locomotion is called tentacle.
Polyp	Small, abnormal growth that arises from the epithelial lining is referred to as polyp. In zoology, a polyp is one of two forms of individuals found in many species of cnidarians. They are approximately cylindrical, elongated on the axis of the body.
Nematocyst	In cnidarians, a capsule that contains a threadlike fiber whose release aids in the capture of prey is referred to as nematocyst.
Mucus	Mucus is a slippery secretion of the lining of various membranes in the body (mucous membranes). Mucus aids in the protection of the lungs by trapping foreign particles that enter the nose during normal breathing. Additionally, it prevents tissues from drying out.
Enzyme	An enzyme is a protein that catalyzes, or speeds up, a chemical reaction. They are essential to sustain life because most chemical reactions in biological cells would occur too slowly, or would lead to different products, without them.
Epidermis	Epidermis is the outermost layer of the skin. It forms the waterproof, protective wrap over the body's surface and is made up of stratified squamous epithelium with an underlying basement membrane. It contains no blood vessels, and is nourished by diffusion from the dermis. In plants, the outermost layer of cells covering the leaves and young parts of a plant is the epidermis.
Medusa	Medusa is a form of cnidarian in which the body is shortened on its principal axis and broadened, sometimes greatly, in contrast with the hydroid or polyp.
Water column	The vertical column of seawater that extends from the surface to the bottom is called water column.
Stalk	The stalk is the longish piece that supports the seed-carrying parts of a plant, or, more simply, the stem. A leaf's petiole; the slender stem that supports the blade of a leaf and attaches it to a larger stem of the plant.
Gamete	A gamete is a specialized germ cell that unites with another gamete during fertilization in organisms that reproduce sexually. They are haploid cells; that is, they contain one complete set of chromosomes. When they unite they form a zygote—a cell having two complete sets of chromosomes and therefore diploid.
Anthozoa	Anthozoa is a class within the phylum Cnidaria that contains the Sea anemones and corals. Unlike other cnidarians, anthozoans do not have a medusa stage in their development. Instead, they release sperm and eggs that form a planula, which attaches to some substrate on which the cnidarian grows.
Zygote	Diploid cell formed by the union of sperm and egg is referred to as zygote.
Larva	A free-living, sexually immature form in some animal life cycles that may differ from the adult in morphology, nutrition, and habitat is called larva.
Polymorphism	The presence in a population of more than one allele of a gene at a frequency greater than that of newly arising mutations is referred to as polymorphism.
Snail	The name snail applies to most members of the molluscan class Gastropoda that have coiled shells. Other gastropods, which lack a conspicuous shell, are commonly called slugs, and are scattered throughout groups that primarily include snails.

Crab	The term crab is often applied to several different groups of short (nose to tail) decapod crustaceans with thick exoskeletons, but only members of the Brachyura are true.
Seaweed	Phycologists consider seaweed to refer any of a large number of marine benthic algae that are multicellular, macrothallic (large-bodied), and thus differentiated from most algae that tend to be microscopic in size.
Scyphozoan	Scyphozoan of the phylum Cnidaria is any of various usually free-swimming marine coelenterates having a gelatinous medusoid stage as the dominant phase of its life cycle
Embryo	Embryo refers to a developing stage of a multicellular organism. In humans, the stage in the development of offspring from the first division of the zygote until body structures begin to appear, about the ninth week of gestation.
Carnivore	An animal that eats a diet consisting solely of meat is referred to as a carnivore.
Muscle	Muscle is a contractile form of tissue. It is one of the four major tissue types, the other three being epithelium, connective tissue and nervous tissue. Muscle contraction is used to move parts of the body, as well as to move substances within the body.
Fertilization	Fertilization is fusion of gametes to form a new organism. In animals, the process involves a sperm fusing with an ovum, which eventually leads to the development of an embryo.
Bilateral symmetry	In bilateral symmetry only one plane, called the sagittal plane, will divide an organism into roughly mirror image halves. Thus there is approximate reflection symmetry. Often the two halves can meaningfully be referred to as the right and left halves.
Radial symmetry	Organisms with radial symmetry have body parts arranged in a regular, repeating pattern around a central axis (like a wagon wheel) or are completely symmetrical about a central axis (like a dinner plate). These organisms resemble a pie where several cutting planes produce roughly identical pieces. An organism with radial symmetry exhibits no left or right sides. They have a top and a bottom only.
Fission	A means of asexual reproduction whereby a parent separates into two or more genetically identical individuals of about equal size is referred to as fission.
Soft coral	Soft coral is a group of coral species that are thick and fleshy, and that move and wave in ocean water currents. These corals do not produce calcium carbonate skeleton; they include leather coral, cauliflower coral, and tree coral.
Flatworm	The flatworm is a phylum of relatively simple soft-bodied invertebrate animals. With about 25,000 known species they are the largest phylum of acoelomates. They are found in marine, freshwater, and even damp terrestrial environments.
Ventral	The surface or side of the body normally oriented upwards, away from the pull of gravity, is the dorsal side; the opposite side, typically the one closest to the ground when walking on all legs, swimming or flying, is the ventral side.
Ectoderm	Ectoderm refers to the outer layer of three embryonic cell layers in a gastrula; forms the skin of the gastrula and gives rise to the epidermis and nervous system in the adult.
Endoderm	Cells migrating inward along the archenteron form the inner layer of the gastrula, which develops into the endoderm.
Central nervous system	The central nervous system comprized of the brain and spinal cord, represents the largest part of the nervous system. Together with the peripheral nervous system, it has a fundamental role in the control of behavior.
Brain	The part of the central nervous system involved in regulating and controlling body activity and interpreting information from the senses transmitted through the nervous system is referred to as the brain.

Crustacean	Member of a group of marine arthropods that contains among others shrimps, crabs, crayfish, and lobsters is called crustacean.
Annelid	Annelid refers to member of a phylum of invertebrates that contains segmented worms, such as the earthworm and the clam worm.
Pharynx	The pharynx is the part of the digestive system and respiratory system of many animals immediately behind the mouth and in front of the esophagus.
Eyespot	Eyespot refers to a simple, lensless eye found in various invertebrates, including flatworms and jellyfish. Eyespots can distinguish light from dark and sometimes the direction of light, but they cannot form an image.
Vertebrate	Vertebrate is a subphylum of chordates, specifically, those with backbones or spinal columns. They started to evolve about 530 million years ago during the Cambrian explosion, which is part of the Cambrian period.
Mollusk	The mollusk is the large and diverse phylum Mollusca, which includes a variety of familiar creatures well-known for their decorative shells or as seafood. These range from tiny snails, clams, and abalone to the octopus, cuttlefish and squid (which are considered the most intelligent invertebrates).
Egg	An egg is the zygote, resulting from fertilization of the ovum. It nourishes and protects the embryo.
Roundworm	A member of the phylum Nematoda is called roundworm. They are one of the most common phyla of animals, with over 20,000 different described species.
Nematoda	Nematoda refers to the phylum that contains the roundworms, or nematodes; characterized by a pseudocoelom, a cylindrical, wormlike body form, and a tough cuticle.
Nemertea	The phylum Nemertea contains the ribbon worms or proboscis worms, which are a group of unsegmented marine invertebrates. There are about 1200 species distributed among two orders: Anopla and Enopla.
Anus	In anatomy, the anus is the external opening of the rectum. Closure is controlled by sphincter muscles. Feces are expelled from the body through the anus during the act of defecation, which is the primary function of the anus.
Ribbon worm	A ribbon worm is a marine invertebrate with a complete digestive tract, a true circulatory system, and a long proboscis to capture prey.
Shoot	In botany, the shoot is one of two primary sections of a plant; the other is the root. The shoot refers to what is generally the upper portion of a plant, and consists of stems, leaves, flowers, and fruits. It is derived from the embryonic epicotyl, the portion of the embryo above the point of attachment to the seed leaves (cotyledons).
Cuticle	Cuticle in animals, a tough, nonliving outer layer of the skin. In plants, a waxy coating on the surface of stems and leaves that helps retain water.
Rods	Rods, are photoreceptor cells in the retina of the eye that can function in less intense light than can the other type of photoreceptor, cone cells.
Intestine	The intestine is the portion of the alimentary canal extending from the stomach to the anus and, in humans and mammals, consists of two segments, the small intestine and the large intestine. The intestine is the part of the body responsible for extracting nutrition from food.
Closed system	An ecosystem that can can exchange heat and work, but not matter, with its surroundings is called a closed system.

Earthworm	Earthworm is the common name for the larger members of the Oligochaeta (which is either a class or subclass depending on the author) in the phylum Annelida.
Leech	The leech is a annelid comprising the subclass Hirudinea. There are freshwater, terrestrial and marine leeches. Like their near relatives, the Oligochaeta, they share the presence of a clitellum.Like earthworms, leeches are hermaphrodites.
Digestive gland	An enzyme-producing gland in several groups of invertebrates where digestion and absorption take place is referred to as a digestive gland.
Hydrostatic skeleton	A hydrostatic skeleton or hydroskeleton is a structure found in many soft-bodied invertebrates consisting of a fluid-filled cavity surrounded by muscles. The pressure of the fluid and action of the surrounding muscles are used to change an organism's shape and produce movement.
Polychaeta	The Polychaeta are a class of annelid worms, generally marine, with a pair of fleshy protrusions on each body segment called parapodia that bear many bristles, called setae, which are made of Chitin.
Gill	An extension of the body surface of an aquatic animal, specialized for gas exchange and/or suspension feeding is called a gill.
Salt	Salt is a term used for ionic compounds composed of positively charged cations and negatively charged anions, so that the product is neutral and without a net charge.
Muscle contraction	A muscle contraction occurs when a muscle cell (called a muscle fiber) shortens. There are three general types: skeletal, heart, and smooth.
Parasite	A parasite is an organism that spends a significant portion of its life in or on the living tissue of a host organism and which causes harm to the host without immediately killing it. They also commonly show highly specialized adaptations allowing them to exploit host resources.
Substrate	A substrate is a molecule which is acted upon by an enzyme. Each enzyme recognizes only the specific substrate of the reaction it catalyzes. A surface in or on which an organism lives.
Oligochaete	The Oligochaete or "few-bristled" worms are well-segmented Annelids, most with a spacious coelom that is used as a hydroskeleton. Their setae (chaetae) or "bristles" are generally few in number and they lack the parapodia of the polychaeta.
Micrograph	A micrograph is a photograph or similar image taken through a microscope or similar device to show a magnified image of an item.
Electron	The electron is a light fundamental subatomic particle that carries a negative electric charge. The electron is a spin-1/2 lepton, does not participate in strong interactions and has no substructure.
Capillary	A capillary is the smallest of a body's blood vessels, measuring 5-10 micro meters. They connect arteries and veins, and most closely interact with tissues. Their walls are composed of a single layer of cells, the endothelium. This layer is so thin that molecules such as oxygen, water and lipids can pass through them by diffusion and enter the tissues.
Plankton	Plankton are drifting organisms that inhabit the water column of oceans, seas, and bodies of fresh water. Plankton abundance and distribution are strongly dependent on factors such as ambient nutrients concentrations, the physical state of the water column, and the abundance of other plankton.
Frond	A frond is the leaf structure of a fern. The term is colloquially applied to the leaves of palms, cycads, and other plants with pinnately compound leaves. A significant difference is that, unlike the leaves of the latter, fern fronds bear the reproductive structures (spore-

	bearing structures) of the sporophyte plant.
Phytoplankton	Phytoplankton refers to the autotrophic component of the plankton that drifts in the water column. Phytoplankton obtain energy through a process called photosynthesis, and so must live in the well-lit surface layer (termed the euphotic zone) of an ocean, sea, or lake.
Blood vessel	A blood vessel is a part of the circulatory system and function to transport blood throughout the body. The most important types, arteries and veins, are so termed because they carry blood away from or towards the heart, respectively.
Dorsal	In anatomy, the dorsal is the side in which the backbone is located. This is usually the top of an animal, although in humans it refers to the back.
Nutrition	Nutrition refers to collectively, the processes involved in taking in, assimilating, and utilizing nutrients.
Hemoglobin	Hemoglobin is the iron-containing oxygen-transport metalloprotein in the red cells of the blood in mammals and other animals. Hemoglobin transports oxygen from the lungs to the rest of the body, such as to the muscles, where it releases the oxygen load.
Sulfide	One of the minerals that is abundant in the hot water that seeps through hydrothermal vents is sulfide.
Trochophore	The free-swimming larva of some annelids and mollusks, distinguished by a wheel-like band of cilia around the middle, and indicating an evolutionary relationship between these two groups is a trochophore.
Mantle cavity	A mantle skirt is a double fold of mantle that encloses a water space. This space is called the mantle cavity, and it is a central feature of mollusk biology, containing the mollusk's gills, anus, osphradium, nephridiopores, and gonopores.
Respiration	Respiration is the process by which an organism obtains energy by reacting oxygen with glucose to give water, carbon dioxide and ATP (energy). Respiration takes place on a cellular level in the mitochondria of the cells and provide the cells with energy.
Mantle	The mantle is an organ found in mollusks. It is the dorsal body wall covering the main body, or visceral mass. The epidermis of this organ secretes calcium carbonate to create a shell.
Mussel	A mussel is a bivalve mollusk that can be found in lakes, rivers, creeks, intertidal areas, and throughout the ocean. The saltwater mussels (family Mytilidae) and freshwater mussels (family Unionidae) are not thought to be closely related and are grouped in different subclasses, despite considerable similarities in appearance.
Chiton	A mollusc that has a shell divided into eight overlapping plates is a chiton.
Bivalve	Bivalve refers to a member of a group of mollusks that includes clams, mussels, scallops, and oysters. They typically have two-part shells, with both parts being more or less symmetrical.
Byssal thread	Byssal thread denotes strong threads secreted by mussels to attach to rocks and large, generally heavy objects in the intertidal zone. They range to 6 centimeters in length.
Digestive system	The organ system that ingests food, breaks it down into smaller chemical units, and absorbs the nutrient molecules is referred to as the digestive system.
Siphon	The tube-like extension through which water flows in and out of the mantle cavity in bivalves and cephalopods, and in tunicates is called siphon.
Nut	A nut in botany is a simple dry fruit with one seed (rarely two) in which the ovary wall becomes very hard (stony or woody) at maturity, and where the seed remains unattached or unfused with the ovary wall.
Cellulase	Cellulase refers to an enzyme that catalyzes the breakdown of the carbohydrate cellulose into

	its component glucose molecules; almost entirely restricted to microorganisms.
Nitrogen	A colorless and tasteless and mostly inert diatomic non-metal gas that is an essential constituent of proteins is nitrogen.
Gland	A gland is an organ in an animal's body that synthesizes a substance for release such as hormones, often into the bloodstream or into cavities inside the body or its outer surface.
Acid	An acid is a water-soluble, sour-tasting chemical compound that when dissolved in water, gives a solution with a pH of less than 7.
Radula	A toothed, grasping organ found in many mollusks, used to scrape up or shred food is referred to as a radula.
Gastropod	A member of the largest group of mollusks, including snails and slugs is a gastropod.
Operculum	The operculum is partly in the most posterior portion of the inferior frontal gyrus of the frontal lobe in the brain. One famous part of the operculum is Broca's area which plays an important role in converzation or speech production, reading and writing.
Limpet	Limpet is a marine mollusk in the family Acmaeidae with flattened, cone-shaped shells. They live throughout the intertidal zone, from the high zone (upper littoral) to the shallow subtidal on the rocky coasts of most oceans. They can be commonly found attached to rocks, looking like little disks or bumps on the rock surface.
Whelk	A whelk is a large marine gastropod (snail) found in temperate waters. Whelks are different from, but sometimes confused as, conches (pronounced "Konks"). They are scavengers and carnivores, equipped with an extensible proboscis that is tipped with a file-like radula.
Planktivore	An animal that feeds mostly on plankton is the planktivore.
Herbivore	A herbivore is an animal that is adapted to eat primarily plant matter
Trait	In biology, a trait or character is a genetically inherited feature of an organism: Eye color is a character; brown eyes and blue eyes is a trait.
Lacuna	Small pit or hollow cavity, as in bone or cartilage, where a cell or cells are located is referred to as the lacuna.
Cephalopod	Cephalopod refers to a member of a group of mollusks that includes squids and octopuses, characterized by bilateral body symmetry.
Diaphragm	The diaphragm is a shelf of muscle extending across the bottom of the ribcage. It is critically important in respiration: in order to draw air into the lungs, the diaphragm contracts, thus enlarging the thoracic cavity and reducing intra-thoracic pressure.
Retina	The retina is a thin layer of cells at the back of the eyeball of vertebrates and some cephalopods; it is the part of the eye which converts light into nervous signals.
Cornea	The cornea is the transparent front part of the eye that covers the iris, pupil, and anterior chamber and provides most of an eye's optical power.
Iris	The colored part of the vertebrate eye, formed by the anterior portion of the choroid is called the iris.
Lens	The lens or crystalline lens is a transparent, biconvex structure in the eye that, along with the cornea, helps to refract light to focus on the retina. Its function is thus similar to a man-made optical lens.
Eye	An eye is an organ that detects light. Different kinds of light-sensitive organs are found in a variety of creatures.
Tusk shell	The tusk shell is a class Scaphopoda of marine mollusks distinguished by openly curved

	planispiral tubular shells open at both ends.
Compound eye	Compound eye refers to a type of eye, found in arthropods, that is composed of numerous independent subunits called ommatidia. Each ommatidium apparently contributes a piece of a mosaiclike image perceived by the animal.
Arthropod	Arthropod refers to member of a phylum of invertebrates that contains among other groups crustaceans and insects that have an exoskeleton and jointed appendages.
Chitin	Chitin refers to strong but flexible nitrogenous polysaccharide found in fungal cells and in the exoskeleton of arthropods.
Centipede	A carnivorous terrestrial arthropod that has one pair of long legs for each of its numerous body segments, with the front pair modified as poison claws is a centipede.
Millipede	A terrestrial arthropod that has two pairs of short legs for each of its numerous body segments and that eats decaying plant matter is referred to as the millipede.
Arachnid	The arachnid is a class of joint-legged invertebrate animals in the subphylum Chelicerata. They are chiefly terrestrial arthropods, some 65,000 to 73,000 species including the spiders, scorpions, harvestmen, ticks, and mites. They are easily distinguished from the insects by the fact that they have eight legs instead of six.
Insect	An arthropod that usually has three body segments , three pairs of legs, and one or two pairs of wings is called an insect. They are the largest and (on land) most widely-distributed taxon within the phylum Arthropoda. They comprise the most diverse group of animals on the earth, with around 925,000 species described
Crustacea	The animal phylum that includes mostly aquatic animals such as crabs, lobsters, crayfish, and shrimp is referred to as crustacea.
Subphylum	A subphylum is a taxonomic rank intermediate between phylum and superclass.
Polysaccharide	Polymer made from sugar monomers is a polysaccharide. They are relatively complex carbohydrates.
Extinction	In biology and ecology, extinction is the ceasing of existence of a species or group of taxa. The moment of extinction is generally considered to be the death of the last individual of that species.The death of all members of a species is extinction.
Cambrian	The Cambrian is a major division of the geologic timescale that begins about 542 Ma (million years ago) at the end of the Proterozoic eon and ended about 488.3 Ma with the beginning of the Ordovician period (ICS, 2004).
Permian	The Permian is a geologic period that extends from about 299.0 Ma to 248.0 Ma. It is the last period of the Palaeozoic Era. As with most older geologic periods, the strata that define the Permian are well identified, but the exact date of the period's start is uncertain by a few million years.
Cephalothorax	The anterior portion of the body of many arachnid and malacostracan arthropods, which consists of the head fused with the trunk is a cephalothorax.
Abdomen	The abdomen is a part of the body. In humans, and in many other vertebrates, it is the region between the thorax and the pelvis. In fully developed insects, the abdomen is the third (or posterior) segment, after the head and thorax.
Thorax	The thorax is a division of an animal's body that lies between the head and the abdomen. In humans, the thorax is the region of the body that extends from the neck to the diaphragm, not including the upper limbs.
Molting	In animals, molting is the routine shedding off old feathers in birds, or of old skin in

	reptiles, or of old hairs in mammals. In arthropods, such as insects, arachnids and crustaceans, molting describes the shedding of its exoskeleton (which is often called its shell), typically to let it grow.
Metamorphosis	Metamorphosis is a process in biology by which an individual physically develops after birth or hatching, and involves significant change in form as well as growth and differentiation.
Sensory nerves	Sensory nerves bring impulses toward the central nervous system.
Antennae	Antennae are paired appendages connected to the front-most segments of arthropods. In crustaceans, they are biramous and present on the first two segments of the head, with the smaller pair known as antennules. In insects, olfactory receptors on the antennae bind to odor molecules, including pheromones.
Receptor	A receptor is a protein on the cell membrane or within the cytoplasm or cell nucleus that binds to a specific molecule (a ligand), such as a neurotransmitter, hormone, or other substance, and initiates the cellular response to the ligand. Receptor, in immunology, the region of an antibody which shows recognition of an antigen.
Mosaic	An organism containing tissues of different genotypes is referred to as mosaic.
Trilobite	Trilobite refers to members of an extinct phylum of arthropods with pronounced segmentation.
Telson	The telson is the last division of the body of a crustacean. It is not considered a true segment because it never carries any appendages. Together with the uropods, it forms the tail fan of lobsters, shrimp and other decapods.
Paleozoic	The Paleozoic Era is a major division of the geologic timescale, one of four geologic eras. The division of time into eras, the largest division of geologic time, dates back to Giovanni Arduino in the 18th century, although his original name for the era now called the Paleozoic was "Primitive".
Horseshoe crab	Horseshoe crab refers to a bottom-dwelling arthropod that is more closely related to spiders than crabs, that belongs to the phylum Arthropoda.
Chelicerata	The animal phylum that includes horseshoe crabs, scorpions, ticks, spiders, and an extinct group called the eurypterids is called chelicerata.
Sea spiders	Marine arthropods that have a reduced body and four pairs of legs are called sea spiders.
Merostomata	Merostomata is a class of marine arthropods which includes horseshoe crabs and eurypterids. It includes only four living species.
Chelicerae	Clawlike feeding appendages characteristic of the chelicerate group is a chelicerae.
Dorsal surface	The dorsal surface is arched from above downward, and is subdivided into two unequal parts by the spine; the portion above the spine is called the supraspinatous fossa, and that below it the infraspinous fossa.
Mudflat	Mudflat is a coastal wetland that forms when mud is deposited by the tides or rivers, they are found in sheltered areas such as bays, bayous, lagoons, and estuaries.
Copepods	A group of small crustaceans that are important members of marine and freshwater plankton communities is copepods.
Copepod	The group of small crustaceans that is an important member of marine and freshwater plankton communities is copepod. They are very important food organisms for small fish, whales, and other crustaceans such as krill in the ocean and in fresh water.
Downstream	Downstream refers to a relative position in DNA or RNA. Relative to the position on the strand, downstream is the region towards the 3' end of the strand.

Tide	The periodic, rhythmic rise and fall of the sea surface caused by changes in gravitational forces external to the Earth is referred to as the tide.
Amphipods	Amphipods include over 7000 described species of small, shrimp-like crustaceans. Most amphipods are marine, although a few live in freshwater or are terrestrial. One species common to mid-Atlantic and estuarine waters is Jassa falcata.
Isopod	Isopod is one of the most diverse orders of Crustaceans, with many species living in all environments, and are common in shallow marine waters. Unlike most crustaceans, isopods are successful on land (suborder Oniscidea, woodlice), although their greatest diversity remains in the deep sea (suborder Asellota).
Migration	Migration occurs when living things move from one biome to another. In most cases organisms migrate to avoid local shortages of food, usually caused by winter. Animals may also migrate to a certain location to breed, as is the case with some fish.
Estuary	An estuary is a semi-enclosed coastal body of water which has a free connection with the open sea and within which sea water mixes with fresh water. The key feature of an estuary is that it is a mixing place for sea water and a stream or river to supply fresh water.
Barnacle	A barnacle is a type of arthropod belonging to infraclass Cirripedia in the subphylum Crustacea and is hence distantly related to crabs and lobsters. They live attached to surfaces and are typically enclosed by heavy calcareous plates are referred to as barnacles.
Sessile	In botany, sessile means without a stalk, as in flowers or leaves that grow directly from the stem. In zoology, sessile organisms are those which are not able to move about. They are usually permanently attached to a solid substrate of some kind, such as a rock, or the hull of a ship in the case of barnacles.
Lophophore	The lophophore is a characteristic feeding organ possessed by three major groups of animals: the Brachiopoda, Bryozoa, and Phoronida. It can most easily be described as a ring of ciliated tentacles surrounding the mouth, but it is often horseshoe-shaped or coiled.
Convergent evolution	An evolutionary pattern in which widely different organisms show similar characteristics because of similar ecological roles and selection pressures is called convergent evolution.
Ganglion	In vertebrate anatomy, a ganglion is a tissue mass that contains the dendrites and cell bodies (or "somata") of nerve cells, in most case ones belonging to the peripheral nervous system.
Zooid	One of the distinct individuals forming a colonial animal such as a bryozoan or hydrozoan is called zooid.
Moss	Moss is a small plant rarely taller than 2 inches (50 mm). They typically grow close together in clumps or mats in damp or shady locations. They do not have flowers and their simple leaves cover the thin wiry stems. At certain times they produces spore capsules which may appear as beak-like capsules borne aloft on thin stalks.
Bryozoan	Colonial animals that build stony skeletons of calcium carbonate and superficially resemble coral is referred to as bryozoan.
Stolon	A stolon is an aerial shoot from a plant with the ability to produce adventitious roots and new offshoots of the same plant.
Runner	A horizontally growing stem that may develop new plants at nodes that touch the soil is called runner.
Brachiopod	A brachiopod is an animala that superficially resembles clams and other bivalve mollusks, but the two halves of the brachiopod shell are dorsal and ventral to the animal rather than lateral, as in clams.

Fossil	A preserved remnant or impression of an organism that lived in the past is referred to as fossil.
Neuron	The neuron is a major class of cells in the nervous system. In vertebrates, they are found in the brain, the spinal cord and in the nerves and ganglia of the peripheral nervous system, and their primary role is to process and transmit neural information.
Substratum	Substratum is the underlying layer, or base to which a lichen is fixed.
Phoronids	Tube-dwelling marine worms commonly known as horseshoe worms, ranging from 1 mm to 50 cm in length are referred to as phoronids.
Echinodermata	Echinodermata refers to the phylum of echinoderms, including sea stars, sea urchins, and sand dollars; characterized by a rough or spiny skin, a water vascular system, an endoskeleton, and radial symmetry in adults.
Echinoderm	Echinoderm is a phylum of marine animals found in the ocean at all depths. This phylum dates back to the lower Cambrian period and represents about 7000 living species and 13000 extinct ones. 6 classes made it to the Modern Era.
Ossicle	An ossicle is one of the small bones of the middle ear. They serve to transmit sounds from the air to the fluid filled labyrinth (cochlea). They are in order from the eardrum to the inner ear, the malleus, incus, and stapes.
Sea cucumber	The sea cucumber is an echinoderm of the class Holothuroidea, with an elongated body and leathery skin. Mostly found on the sea floor.
Brittle star	A brittle star is an echinoderm with five flexible arms that radiate from a conspicuous central disk and tube feet that are used in feeding.
Sea urchin	A sea urchin is a spiny sea creature of the class Echinoidea found in oceans all over the world. Their shell, which biologists call the test, is globular in shape, and covered with spines.
Echinoidea	Sea urchins are spiny sea creatures of the class Echinoidea found in oceans all over the world. (The name sea urchin means sea hedgehog, hedgehog being one meaning of the word urchin). Their shell, which biologists call the test, is globular in shape, and covered with spines. The size of the test in adults is typically in the range of 3 to 10 cm.
Crinoidea	The class Crinoidea of the echinoderms live both in shallow water and in depths as great as 6000 meters. They are characterized by a mouth on the top surface that is surrounded by feeding arms.
Starfish	Starfish are marine invertebrates belonging to phylum Echinodermata, class Asteroidea. They exhibit a superficially radial symmetry, typically with five or more "arms" which radiate from an indistinct disk. They do not have movable skeletons, but instead possess a hydraulic water vascular system.
Crinoids	Crinoids refers to the echinoderms with a small, cup-shaped body and feathery arms: the sea lilies and feather stars .
Reproductive system	A reproductive system is the ensembles and interactions of organs and or substances within an organism that stricly pertain to reproduction. As an example, this would include in the case of female mammals, the hormone estrogen, the womb and eggs but not the breast.
Tube feet	Tube feet are the many small tubular projections found most famously on the ventral face of a sea star's arms, but are characteristic of the water vascular system of the echinoderm phylum which also includes sea urchins, sand dollars and sea cucumbers and many other sea creatures.
Vascular	In botany vascular refers to tissues that contain vessels for transporting liquids. In anatomy and physiology, vascular means related to blood vessels, which are part of the

	Circulatory system.
Madreporite	Small red or yellow button-like structure that often looks like a small wart on a sea star's central disk. The madreporite is used to filter water into the sea star's water-vascular system.
Sieve plate	Perforated wall area in a sieve-tube element through which strands connecting sieve-tube protoplasts pass is a sieve plate.
Ampulla	Base of a semicircular canal in the inner ear is called ampulla. The Ampulla also refers to a dilated segment in a tubular structure, and they are also the bulb-like structures above the tube feet in echinoderms.
Bulb	An underground food-storage organ that is essentially a modified bud consisting of fleshy leaves that surround and are attached to a small stem is called bulb.
Stomach	The stomach is an organ in the alimentary canal used to digest food. It's primary function is not the absorption of nutrients from digested food; rather, the main job of the stomach is to break down large food molecules into smaller ones, so that they can be absorbed into the blood more easily.
Heart urchin	A burrowing sea urchin with a heart-shaped or oval body, and a flattened modified spine is a heart urchin.
Sand dollar	The common sand dollar is the skeleton or test of a marine animal. By the time the test washes up on the beach, it is missing its velvety covering of minute spines and has a somewhat bleached appearance due to its exposure to the sun.
Diadema antillarum	Diadema antillarum is a sea urchin of the genus diadema with long black spines, it was once the most abundant and important herbivore on the coral reefs of the western Atlantic and Caribbean basin. In this role it was the main grazer that maintained the balance between algae growth and destruction.
Ligament	A ligament is a short band of tough fibrous connective tissue composed mainly of long, stringy collagen fibres. They connect bones to other bones to form a joint. (They do not connect muscles to bones.)
Secretion	Secretion is the process of segregating, elaborating, and releasing chemicals from a cell, or a secreted chemical substance or amount of substance.
Zooplankton	Nonphotosynthetic protists that feed on other plankton and are abundant in marine and freshwater environments are zooplankton.
Respiratory tree	Respiratory tree is the branching portion of the respiratory tract within the thoracic cavity.
Commensal	Living on or within another organism without injuring or benefiting the other organism is referred to as commensal.
Sea stars	Echinoderms with five or more radiating arms and tube feet that are used in locomotion are sea stars.
Sea squirts	Tunicates with a sac-like, attached body called a tunic as adults are referred to as sea squirts.
Cellulose	A large polysaccharide composed of many glucose monomers linked into cable-like fibrils that provide structural support in plant cell walls is referred to as cellulose.
Sulfuric acid	Sulfuric acid refers to the major constituent of acid precipitation. Formed as a result of sulfur dioxide emissions reacting with water vapor in the atmosphere.
Heavy metals	The heavy metals are a group of elements between copper and lead on the periodic table of the

	elements—having atomic weights between 63.546 and 200.590 and specific gravities greater than
Predation	Interaction in which one organism uses another, called the prey, as a food source is referred to as predation.
Notochord	The notochord is a flexible rod-shaped body found in embryos of all chordates. It is composed of cells derived from the mesoblast and defining the primitive axis of the embryo. In lower vertebrates, it persists throughout life as the main axial support of the body, while in higher vertebrates it is replaced by the vertebral column.
Lancelet	The lancelet is a group of primitive chordates. They are an important object of study in zoology as they provide indications about the origins of the vertebrates.
Tadpole	A tadpole is a larval frog, toad, salamander, newt, or caecilian. In this stage it breathes by means of external or internal gills, is at first lacking legs, and has a finlike tail with which it swims as most fish do, by lateral undulation.
Tide pool	A depression that holds seawater at low tide is referred to as a tide pool.
Biology	Biology is the branch of science dealing with the study of life. It is concerned with the characteristics, classification, and behaviors of organisms, how species come into existence, and the interactions they have with each other and with the environment.
Embryology	Embryology is the branch of developmental biology that studies embryos and their development
Deuterostome	Deuterostome is a subtaxon of the subregnum Bilateria, and is opposed to the protostomes. They are distinguished by their embryonic development; in a deuterostome, the first opening (the blastopore) becomes the anus, while in protostomes it becomes the mouth.
Protostome	A protostome is a group of coelomate animals in which the first embryonic opening is associated with the mouth. In protostome development, the mouth forms at the site of the blastopore, and the anus forms as a second opening.
Cleavage	Cleavage refers to cytokinesis in animal cells and in some protists, characterized by pinching in of the plasma membrane.
Gastrula	The embryonic stage resulting from gastrulation in animal development. Most animals have a gastrula made up of three layers of cells: ectoderm, endoderm, and mesoderm.
Blastula	The blastula is an early stage of embryonic development in animals. It is produced by cleavage of a fertilized ovum and consisting of a spherical layer of cells surrounding a fluid-filled cavity.
Blastopore	A blastopore is an opening into the archenteron during the embryonic stages of an organism. The distinction between protostomes and deuterostomes is based on the direction in which the mouth (stoma) develops in relation to the blastopore.
Evolutionary tree	The evolutionary tree of living things is currently based on ideas from cladistics; where more than two groups are shown in a single branch, there is disagreement about how they diverged. Hypothetically taxonomy would follow the tree whenever possible, but in many places it does not at present.
Multicellular	Multicellular organisms are those organisms consisting of more than one cell, and having differentiated cells that perform specialized functions. Most life that can be seen with the naked eye is multicellular, as are all animals (i.e. members of the kingdom Animalia) and plants (i.e. members of the kingdom Plantae).
Budding	A means of asexual reproduction whereby a new individual developed from an outgrowth of a parent splits off and lives independently is referred to as budding.
Epithelium	Epithelium is a tissue composed of a layer of cells. Epithelium can be found lining internal

	(e.g. endothelium, which lines the inside of blood vessels) or external (e.g. skin) free surfaces of the body. Functions include secretion, absorption and protection.
Comparative embryology	The study of the formation, early growth, and development of different organisms is called comparative embryology. As they grow, the differences that will distinguish the embryos as adults become more and more apparent.
DNA sequencing	The process of determining the chemical composition of a DNA molecule is called DNA sequencing.
Radioactive dating	Radioactive dating refers to a method of determining the age of fossils and rocks using half-lives of radioactive isotopes.
Big bang theory	The Big Bang theory is the scientific theory that the universe emerged from an enormously dense and hot state about 13.7 billion years ago.
Fauna	Fauna (animals), a collective term for animal life, as distinct from flora (plant life).
Molecular evolution	Molecular evolution refers to changes in protein and DNA sequences over time. We use this information to estimate how recently species diverged from a common ancestor.
DNA sequence	A DNA sequence is a succession of letters representing the primary structure of a real or hypothetical DNA molecule or strand, The possible letters are A, C, G, and T, representing the four nucleotide subunits of a DNA strand (adenine, cytosine, guanine, thymine).
Molecular clock	The molecular clock is a technique in genetics, which researchers use to date when two species diverged. It deduces elapsed time from the number of minor differences between their DNA sequences.
Cambrian explosion	Cambrian explosion refers to a burst of evolutionary origins when most of the major body plans of animals appeared in a relatively brief time in geologic history; recorded in the fossil record about 545 to 525 million years ago.
Fossil record	The chronicle of evolution over millions of years of geologic time engraved in the order in which fossils appear in rock strata is called a fossil record.
Squamata	Squamata is the largest recent order of reptiles, including lizards and snakes. Members of the order are distinguished by their skins, which bear horny scales or shields.
Phylogeny	The science that explores the evolutionary relationships among organisms and seeks to reconstruct evolutionary history is phylogeny.
Reproduction	Biological reproduction is the biological process by which new individual organisms are produced. Reproduction is a fundamental feature of all known life; each individual organism exists as the result of reproduction by an antecedent.
Ecology	Ecology is the scientific study of the distribution and abundance of living organisms and how these properties are affected by interactions between the organisms and their environment.
Monophyletic	A group is monophyletic if it consists of a common ancestor and all its descendants. A taxonomic group that contain organisms but not their common ancestor is called polyphyletic, and a group that contains some but not all descendants of the most recent common ancestor is called paraphyletic.
Ribosomal RNA	The most prominent examples of non-coding Ribosomal RNA are non-coding RNA that are involved in the process of translation and gene expression.
Animal kingdom	Animal kingdom is one of the main divisions or life-waves of entities on earth, separated from the human kingdom by its lack of the emanated or evolved self-conscious mind
Precambrian	The Precambrian is an informal name for the eons of the geologic timescale that came before the current Phanerozoic eon. It spans from the formation of Earth around 4500 Ma (million

	years ago) to the evolution of abundant macroscopic hard-shelled fossils, which marked the beginning of the Cambrian.
Paleontology	Paleontology refers to the scientific study of fossils and all aspects of extinct life.
Genetics	Genetics is the science of genes, heredity, and the variation of organisms.

Microorganism	A microorganism is an organism that is so small that it is microscopic (invisible to the naked eye). They are often illustrated using single-celled, or unicellular organisms; however, some unicellular protists are visible to the naked eye, and some multicellular species are microscopic.
Seaweed	Phycologists consider seaweed to refer any of a large number of marine benthic algae that are multicellular, macrothallic (large-bodied), and thus differentiated from most algae that tend to be microscopic in size.
Crab	The term crab is often applied to several different groups of short (nose to tail) decapod crustaceans with thick exoskeletons, but only members of the Brachyura are true.
Kelp	Kelp are large seaweeds, belonging to the brown algae and classified in the order Laminariales. Despite their appearance they are not grouped with the normal aquatic or land plants (kingdom Plantae), but instead are included in either kingdom Protista or Chromista.
Grazing	Grazing refers to consumption of a portion of a plant's tissues, generallyapplied to grasses and other herbaceous vegetation; removing plant or algal cover from a substrate, generally applied to aquatic systems.
Snail	The name snail applies to most members of the molluscan class Gastropoda that have coiled shells. Other gastropods, which lack a conspicuous shell, are commonly called slugs, and are scattered throughout groups that primarily include snails.
Bacteria	The domain that contains procaryotic cells with primarily diacyl glycerol diesters in their membranes and with bacterial rRNA. Bacteria also is a general term for organisms that are composed of procaryotic cells and are not multicellular.
Decomposition	Decomposition refers to the reduction of the body of a formerly living organism into simpler forms of matter.
Nutrition	Nutrition refers to collectively, the processes involved in taking in, assimilating, and utilizing nutrients.
Photosynthesis	Photosynthesis is a biochemical process in which plants, algae, and some bacteria harness the energy of light to produce food. Ultimately, nearly all living things depend on energy produced from photosynthesis for their nourishment, making it vital to life on Earth.
Phosphate	A phosphate is a polyatomic ion or radical consisting of one phosphorus atom and four oxygen. In the ionic form, it carries a -3 formal charge, and is denoted PO_4^{3-}.
Nitrate	Nitrate refers to a salt of nitric acid; a compound containing the radical NO_3; biologically, the final form of nitrogen from the oxidation of organic nitrogen compounds.
Sugar	A sugar is the simplest molecule that can be identified as a carbohydrate. These include monosaccharides and disaccharides, trisaccharides and the oligosaccharides. The term "glyco-" indicates the presence of a sugar in an otherwise non-carbohydrate substance.
Carbon dioxide	Carbon dioxide is an atmospheric gas comprized of one carbon and two oxygen atoms. A very widely known chemical compound, it is frequently called by its formula CO_2. In its solid state, it is commonly known as dry ice.
Prokaryote	A prokaryote is an organisms without a cell nucleus, or indeed any other membrane-bound organelles, in most cases unicellular (in rare cases, multicellular). This is in contrast to a eukaryote, organisms that have cell nuclei and may be variously unicellular or multicellular.
Eukaryote	A eukaryote is an organism with complex cells, in which the genetic material is organized into membrane-bound nuclei. They comprise animals, plants, and fungi—which are mostly multicellular—as well as various other groups that are collectively classified as protists.

Mitochondria	Mitochondria are organelles found in most eukaryotic cells, including those of plants, animals, fungi, and protists. Mitochondria are sometimes described as "cellular power plants", because their primary function is to convert organic materials into energy in the form of ATP.
Reproduction	Biological reproduction is the biological process by which new individual organisms are produced. Reproduction is a fundamental feature of all known life; each individual organism exists as the result of reproduction by an antecedent.
Chloroplast	A chloroplast is an organelle found in plant cells and eukaryotic algae which conduct photosynthesis. They are similar to mitochondria but are found only in plants. They are surrounded by a double membrane with an intermembrane space; they have their own DNA and are involved in energy metabolism;
Organelle	In cell biology, an organelle is one of several structures with specialized functions, suspended in the cytoplasm of a eukaryotic cell.
Nucleus	In cell biology, the nucleus is found in all eukaryotic cells that contains most of the cell's genetic material. The nucleus has two primary functions: to control chemical reactions within the cytoplasm and to store information needed for cellular division.
Mitosis	Mitosis is the process by which a cell separates its duplicated genome into two identical halves. It is generally followed immediately by cytokinesis which divides the cytoplasm and cell membrane.
Cell	The cell is the structural and functional unit of all living organisms, and is sometimes called the "building block of life."
Gamete	A gamete is a specialized germ cell that unites with another gamete during fertilization in organisms that reproduce sexually. They are haploid cells; that is, they contain one complete set of chromosomes. When they unite they form a zygote—a cell having two complete sets of chromosomes and therefore diploid.
Substratum	Substratum is the underlying layer, or base to which a lichen is fixed.
Morphology	The scientific study of organic form, including both its development and function is morphology.
Adaptation	A biological adaptation is an anatomical structure, physiological process or behavioral trait of an organism that has evolved over a period of time by the process of natural selection such that it increases the expected long-term reproductive success of the organism.
Soil	Soil is material capable of supporting plant life. Soil forms through a variety of soil formation processes, and includes weathered rock "parent material" combined with dead and living organic matter and air.
Shoot	In botany, the shoot is one of two primary sections of a plant; the other is the root. The shoot refers to what is generally the upper portion of a plant, and consists of stems, leaves, flowers, and fruits. It is derived from the embryonic epicotyl, the portion of the embryo above the point of attachment to the seed leaves (cotyledons).
Root	In vascular plants, the root is that organ of a plant body that typically lies below the surface of the soil. However, this is not always the case, since a root can also be aerial (that is, growing above the ground) or aerating (that is, growing up above the ground or especially above water).
Stem	Stem refers to that part of a plant's shoot system that supports the leaves and reproductive structures.
Stomata	A stomata is a tiny opening or pore, found mostly on the undersurface of a plant leaf, and

	used for gas exchange.
Transpiration	Transpiration refers to plant's loss of water to the atmosphere, mainly through evaporation at leaf stomata.
Species	Group of similarly constructed organisms capable of interbreeding and producing fertile offspring is a species.
Tissue	Group of similar cells which perform a common function is called tissue.
Water column	The vertical column of seawater that extends from the surface to the bottom is called water column.
Holdfast	A rootlike structure that anchors a seaweed is called holdfast.
Thallus	Thallus is an undifferentiated vegetative tissue of some non-mobile organisms, which were previously known as the thallophytes.
Stipe	A stipe is a supportive structure that may be stemlike, as in seaweed, or a true leaf stem, as in ferns.
Crust	A crust is the outer layer of a planet, part of its lithosphere. Planetary crust is generally composed of a less dense material than that of its deeper layers. The crust of the Earth is composed mainly of basalt and granite.
Cell division	Cell division is the process by which a cell (called the parent cell) divides into two cells (called daughter cells). Cell division is usually a small segment of a larger cell cycle. In meiosis, however, a cell is permanently transformed and cannot divide again.
Meristem	Meristem is a type of embryonic tissue in plants consisting of unspecialized, youthful cells called meristematic cells and found in areas of the plant where growth is or will take place - the roots and shoots.
Chlorophyll	Chlorophyll is a green photosynthetic pigment found in plants, algae, and cyanobacteria. In plant photosynthesis incoming light is absorbed by chlorophyll and other accessory pigments in the antenna complexes of photosystem I and photosystem II.
Pneumatocyst	A pneumatocyst is a large float containing gas found in brown algae. Plants may have more than one. Gas proportion in the pneumatocyst will adjust composition by equilibrating with the surrounding air or water. The pneumatocyst can hold O_2, CO_2 or CO.
Sargasso sea	Sargasso sea refers to the region of the Atlantic Ocean north of the West Indies that is characterized by floating masses of Sargasso weed, a brown alga.
Habitat	Habitat refers to a place where an organism lives; an environmental situation in which an organism lives.
Pigment	Pigment is any material resulting in color in plant or animal cells which is the result of selective absorption.
Spore	Spore refers to a differentiated, specialized form that can be used for dissemination, for survival of adverse conditions because of its heat and dessication resistance, and/or for reproduction. They are usually unicellular and may develop into vegetative organisms or gametes. A spore may be produced asexually or sexually and are of many types.
Red algae	One of a group of marine, mostly multicellular, autotrophic protists, which includes the reef-building coralline algae is referred to as red algae.
Cell wall	Cell wall refers to a protective layer external to the plasma membrane in plant cells, bacteria, fungi, and some protists; protects the cell and helps maintain its shape.
Cellulose	A large polysaccharide composed of many glucose monomers linked into cable-like fibrils that

	provide structural support in plant cell walls is referred to as cellulose.
Algae	The algae consist of several different groups of living organisms that capture light energy through photosynthesis, converting inorganic substances into simple sugars with the captured energy.
Action spectrum	An action spectrum is the rate of a physiological activity plotted against wavelength of light.
Green algae	Green algae refers to photosynthetic protists that include L.nicellular, colonial, and multicellular species with grass green chloroplasts; closely related to true plants.
Dispersal	Biological dispersal refers to those processes by which a species maintains or expands the distribution of a population. Dispersal implies movement—movement away from an existing population or away from the parent organisms.
Chromosome	A chromosome is, minimally, a very long, continuous piece of DNA, which contains many genes, regulatory elements and other intervening nucleotide sequences.
Haploid	Haploid cells bear one copy of each chromosome.
Diploid	Diploid cells have two copies (homologs) of each chromosome (both sex- and non-sex determining chromosomes), usually one from the mother and one from the father. Most somatic cells (body cells) of complex organisms are diploid.
Gametophyte stage	A life cycle stage in plants in which a haploid sex cell is produced by mitosis is called gametophyte stage.
Brown algae	One of a group of marine, multicellular, autotrophic protists, the most common type of seaweed. Brown algae include the kelps.
Gametangia	In the case of compatibility, gametangia form on the hyphae that are the generative cells for the gametes, in which numerous nuclei gather.
Sporophyte	A sporophyte is the diploid structure or phase of life of a sexually reproducing plant. Each living cell of the sporophyte contains two complete sets of chromosomes. The sporophyte is the dominant life form in ferns, gymnosperms, and angiosperms.
Zygote	Diploid cell formed by the union of sperm and egg is referred to as zygote.
Fusion	Fusion refers to the combination of two atoms into a single atom as a result of a collision, usually accompanied by the release of energy.
Monoecious	Monoecious refers to having unisexual flowers, conifer cones, or functionally equivalent structures of both sexes appearing on the same plant.
Dioecious	A plant having unisexual flowers, conifer cones, or functionally equivalent structures occurring on different individuals is called dioecious.
Gametophyte	Gametophyte refers to the multicellular haploid form in the life cycle of organisms undergoing alternation of generations; mitotically produces haploid gametes that unite and grow into the sporophyte generation.
Herbivore	A herbivore is an animal that is adapted to eat primarily plant matter
Photosynthetic pigment	A photosynthetic pigment is a pigment that is present in chloroplasts or photosynthetic bacteria and captures the light energy necessary for photosynthesis.
Chlorophyta	The Chlorophyta sensu stricto or chlorophyte algae, comprises most of what are commonly called green algae and includes most members of the grade of putatively ancestral scaly flagellates in Prasinophyceae plus members of Ulvophyceae, Trebouxiophyceae, and Chlorophyceae.

Starch	Biochemically, starch is a combination of two polymeric carbohydrates (polysaccharides) called amylose and amylopectin.
Mudflat	Mudflat is a coastal wetland that forms when mud is deposited by the tides or rivers, they are found in sheltered areas such as bays, bayous, lagoons, and estuaries.
Genus	In biology, a genus is a taxonomic grouping. That is, in the classification of living organisms, a genus is considered to be distinct from other such genera. A genus has one or more species: if it has more than one species these are likely to be morphologically more similar than species belonging to different genera.
Xanthophyll	Xanthophyll refers to a yellow or orange pigment commonly found as an accessory pigment in photosynthesis, but found elsewhere as well. An oxygen-containing carotenoid.
Carotene	Carotene is an orange photosynthetic pigment important for photosynthesis. It is responsible for the orange color of the carrot and many other fruits and vegetables. It contributes to photosynthesis by transmitting the light energy it absorbs to chlorophyll.
Phycocolloid	One of several starch-like chemicals found in some seaweeds and are of significant commercial importance is phycocolloid.
Sea urchin	A sea urchin is a spiny sea creature of the class Echinoidea found in oceans all over the world. Their shell, which biologists call the test, is globular in shape, and covered with spines.
Acetone	In chemistry, acetone (also known as propanone, dimethyl ketone, 2-propanone, propan-2-one and â-ketopropane) is the simplest representative of the ketones.
Phycoerythrin	A red photosynthetic phycobiliprotein pigment used to trap light energy is called phycoerythrin.
Filament	The stamen is the male organ of a flower. Each stamen generally has a stalk called the filament, and, on top of the filament, an anther. The filament is a long chain of proteins, such as those found in hair, muscle, or in flagella.
Moss	Moss is a small plant rarely taller than 2 inches (50 mm). They typically grow close together in clumps or mats in damp or shady locations. They do not have flowers and their simple leaves cover the thin wiry stems. At certain times they produces spore capsules which may appear as beak-like capsules borne aloft on thin stalks.
Calcium carbonate	A mineral that is the major component of the shell, skeleton, and other parts of many organisms are called calcium carbonate.
Coralline algae	Red algae that deposit calcium carbonate in their thallus are called the coralline algae.
Coral reef	A coral reef is a type of biotic reef that develops in tropical waters by the growth of coralline algae, hermatypic corals, and other marine organisms.
Agar	Agar is a galactose polymer (or agarose) obtained from the cell walls of some species of red algae or seaweed (Sphaerococcus euchema) and species of Gelidium and Gracilaria, chiefly from eastern Asia, Chile and California.
Flowering plants	The flowering plants (also called angiosperms) are a major group of land plants. They comprise one of the two groups in the seed plants: the flowering plants cover their seeds by including them in a true fruit.
Pollen	The male gametophyte in gymnosperms and angiosperms is referred to as pollen.
Flower	A flower is the reproductive structure of a flowering plant. The flower structure contains the plant's reproductive organs, and its function is to produce seeds through sexual reproduction.

Temperate	In virology, temperate refers to the life cycle a phage is able to perform. The phage usually integrates its genome into its host chromosome, becoming a prophage.
Zostera	Zostera is a small genus of widely distributed seagrass, commonly called eelgrass. It contains twelve species. Zostera is found on sandy sunstrates or in esturaries submerged or partially floating. Most Zostera are perennial, they have long, bright green, ribbon-like leaves, about 1 cm wide.
Stigma	The sticky tip of a flower's carpel, which traps pollen grains is called stigma. In zoology, one of the external openings of the tracheae of insects, myriapods, and other arthropods; a spiracle. One of the apertures of the pulmonary sacs of arachnids. One of the apertures of the gill of an ascidian, and of Amphioxus.
Fruit	A fruit is the ripened ovary—together with seeds—of a flowering plant. In many species, the fruit incorporates the ripened ovary and surrounding tissues.
Population	Group of organisms of the same species occupying a certain area and sharing a common gene pool is referred to as population.
Rhizome	A rhizome is a usually-underground, horizontal stem of a plant that often sends out roots and shoots from its nodes.
Sexual reproduction	The propagation of organisms involving the union of gametes from two parents is sexual reproduction.
Sediment	Sediment is any particulate matter that can be transported by fluid flow and which eventually is deposited as a layer of solid particles on the bed or bottom of a body of water or other liquid.
Symbiotic nitrogen-fixing bacteria	Bacteria that live in the roots of certain kinds of plants, where they convert nitrogen gas molecules into compounds that plants can use is called symbiotic nitrogen-fixing bacteria.
Nitrogen	A colorless and tasteless and mostly inert diatomic non-metal gas that is an essential constituent of proteins is nitrogen.
Residence time	Residence time refers to the ratio of the size of a compartment to the flux through it, expressed in units of time; thus, the average time spent by energy or a substance in the compartment.
Host	Host is an organism that harbors a parasite, mutual partner, or commensal partner; or a cell infected by a virus.
Silica	Silica is a mineral similar to glass that is the major component of the cell wall, shell, or skeleton of many marine organisms.
Diatom	Diatom refers to a unicellular photosynthetic alga with a unique, glassy cell wall containing silica.
Heterotrophic	Incapable of synthesizing food and therefore dependent on other organisms for it is called heterotrophic.
Enzyme	An enzyme is a protein that catalyzes, or speeds up, a chemical reaction. They are essential to sustain life because most chemical reactions in biological cells would occur too slowly, or would lead to different products, without them.
Cyanobacteria	Cyanobacteria are a phylum of bacteria that obtain their energy through photosynthesis. They are often referred to as blue-green algae, even though it is now known that they are not directly related to any of the other algal groups, which are all eukaryotes.
Nitrogen	Nitrogen fixation is the process by which nitrogen is taken from its relatively inert

fixation	molecular form (N_2) in the atmosphere and converted into nitrogen compounds useful for other chemical processes.
Amino acid	An amino acid is any molecule that contains both amino and carboxylic acid functional groups. They are the basic structural building units of proteins. They form short polymer chains called peptides or polypeptides which in turn form structures called proteins.
Ion	Ion refers to an atom or molecule that has gained or lost one or more electrons, thus acquiring an electrical charge.
Trichome	A trichome is a fine outgrowth or appendage on plants and protists. These are of diverse of structure and function. Examples are hairs, glandular hairs, scales, and papillae.
Heterocyst	Heterocyst refers to a large, thick-walled cell in the filaments of certain cyanobacteria; performs nitrogen fixation.
Fixation	Fixation in population genetics occurs when the frequency of a gene reaches 1.Fixation in biochemistry, histology, cell biology and pathology refers to the technique of preserving a specimen for microscopic study, making it intact and stable, but dead.
Oxygen	Oxygen is a chemical element in the periodic table. It has the symbol O and atomic number 8. Oxygen is the second most common element on Earth, composing around 46% of the mass of Earth's crust and 28% of the mass of Earth as a whole, and is the third most common element in the universe.
Interstitial water	The water contained between sediment particles is interstitial water.
Multicellular	Multicellular organisms are those organisms consisting of more than one cell, and having differentiated cells that perform specialized functions. Most life that can be seen with the naked eye is multicellular, as are all animals (i.e. members of the kingdom Animalia) and plants (i.e. members of the kingdom Plantae).
Unicellular	Unicellular organisms carry out all the functions of life. Unicellular species are those whose members consist of a single cell throughout their life cycle. This latter qualification is significant since most multicellular organisms consist of a single cell at the beginning of their life cycles.
Botany	Botany is the scientific study of plant life. As a branch of biology, it is also sometimes referred to as plant science or plant biology.
Microbiology	Microbiology is the study of microorganisms, including viruses, viroids, prions, prokaryotes and simple eukaryotes (e.g. molds and fungi).
Microbe	A microbe is an organism that is so small that it is microscopic (invisible to the naked eye). Microorganisms are often illustrated using single-celled, or unicellular organisms; however, some unicellular protists are visible to the naked eye, and some multicellular species are microscopic.
Salt marsh	Salt marsh refers to a grassy area that extends along the shores of estuaries and sheltered coasts in temperate and subpolar regions.
Seedling	Seedling refers to a young plant that has grown from a seed. The most common example of germination is the sprouting of a seedling from a seed of a flowering plant or gymnosperm.
Mussel	A mussel is a bivalve mollusk that can be found in lakes, rivers, creeks, intertidal areas, and throughout the ocean. The saltwater mussels (family Mytilidae) and freshwater mussels (family Unionidae) are not thought to be closely related and are grouped in different subclasses, despite considerable similarities in appearance.
Mangrove	Mangrove are woody trees or shrubs that grow in coastal habitats or mangal (Hogarth, 1999),

for which the term mangrove swamp also would apply.

Estuary An estuary is a semi-enclosed coastal body of water which has a free connection with the open sea and within which sea water mixes with fresh water. The key feature of an estuary is that it is a mixing place for sea water and a stream or river to supply fresh water.

Fecal Fecal is the solid excretory wastes of animals. Consists of undigested material passing through the gut and bacteria that have begun to feed on it.

Sessile In botany, sessile means without a stalk, as in flowers or leaves that grow directly from the stem. In zoology, sessile organisms are those which are not able to move about. They are usually permanently attached to a solid substrate of some kind, such as a rock, or the hull of a ship in the case of barnacles.

Invertebrate Invertebrate is a term coined by Jean-Baptiste Lamarck to describe any animal without a spinal column. It therefore includes all animals except vertebrates (fish, reptiles, amphibians, birds and mammals).

Barren An area with sparse vegetation owing to some physical or chemical property of the soil is referred to as barren.

Sea anemone Named after a terrestrial flower, the sea anemone is a group of water-dwelling, filter feeding animals of the order Actinaria. As a cnidarian, it is closely related to coral and jellyfish.

Starfish Starfish are marine invertebrates belonging to phylum Echinodermata, class Asteroidea. They exhibit a superficially radial symmetry, typically with five or more "arms" which radiate from an indistinct disk. They do not have movable skeletons, but instead possess a hydraulic water vascular system.

Sea cucumber The sea cucumber is an echinoderm of the class Holothuroidea, with an elongated body and leathery skin. Mostly found on the sea floor.

Deposit feeder A deposit feeder is an organism that feeds on the particles of matter in the soil, usually the top sediment where it is filled with organic matter.

Sponge An invertebrates that consist of a complex aggregation of cells, including collar cells, and has a skeleton of fibers and/or spicules is a sponge. They are primitive, sessile, mostly marine, waterdwelling filter feeders that pump water through their matrix to filter out particulates of food matter.

Colony Colony refers to a cluster or assemblage of microorganisms growing on the surface of an agar culture medium. A colony also refers to several individual organisms of the same species living closely together, usually for mutual benefit, such as stronger defences, the ability to attack bigger prey, etc.

Coral Group of cnidarians having a calcium carbonate skeleton that participate in the formation of reefs is coral.

Tentacle An elongate, extensible projection of the body of cnidarians and cephalopod mollusks that may be used for grasping, stinging, and immobilizing prey, and for locomotion is called tentacle.

Zooplankton Nonphotosynthetic protists that feed on other plankton and are abundant in marine and freshwater environments are zooplankton.

Polyp Small, abnormal growth that arises from the epithelial lining is referred to as polyp. In zoology, a polyp is one of two forms of individuals found in many species of cnidarians. They are approximately cylindrical, elongated on the axis of the body.

Stalk The stalk is the longish piece that supports the seed-carrying parts of a plant, or, more simply, the stem. A leaf's petiole; the slender stem that supports the blade of a leaf and

	attaches it to a larger stem of the plant.
Nudibranchs	Gastropods that lack a shell and have exposed gills are called nudibranchs. They form the largest suborder of the order Opisthobranchia, subclass Orthogastropoda, class Gastropoda in the phylum Mollusca.
Egg	An egg is the zygote, resulting from fertilization of the ovum. It nourishes and protects the embryo.

Benthos	Benthos are the organisms and habitats of the sea floor; in freshwater biology they are the organisms and habitats of the bottoms of lakes, rivers, and creeks.
Habitat	Habitat refers to a place where an organism lives; an environmental situation in which an organism lives.
Water column	The vertical column of seawater that extends from the surface to the bottom is called water column.
Substratum	Substratum is the underlying layer, or base to which a lichen is fixed.
Crinoids	Crinoids refers to the echinoderms with a small, cup-shaped body and feathery arms: the sea lilies and feather stars .
Seaweed	Phycologists consider seaweed to refer any of a large number of marine benthic algae that are multicellular, macrothallic (large-bodied), and thus differentiated from most algae that tend to be microscopic in size.
Limpet	Limpet is a marine mollusk in the family Acmaeidae with flattened, cone-shaped shells. They live throughout the intertidal zone, from the high zone (upper littoral) to the shallow subtidal on the rocky coasts of most oceans. They can be commonly found attached to rocks, looking like little disks or bumps on the rock surface.
Coral	Group of cnidarians having a calcium carbonate skeleton that participate in the formation of reefs is coral.
Sediment	Sediment is any particulate matter that can be transported by fluid flow and which eventually is deposited as a layer of solid particles on the bed or bottom of a body of water or other liquid.
Mussel	A mussel is a bivalve mollusk that can be found in lakes, rivers, creeks, intertidal areas, and throughout the ocean. The saltwater mussels (family Mytilidae) and freshwater mussels (family Unionidae) are not thought to be closely related and are grouped in different subclasses, despite considerable similarities in appearance.
Suspension feeder	An animal that extracts food particles suspended in the surrounding water for food is referred to as a suspension feeder.
Deposit feeder	A deposit feeder is an organism that feeds on the particles of matter in the soil, usually the top sediment where it is filled with organic matter.
Herbivore	A herbivore is an animal that is adapted to eat primarily plant matter
Chiton	A mollusc that has a shell divided into eight overlapping plates is a chiton.
Snail	The name snail applies to most members of the molluscan class Gastropoda that have coiled shells. Other gastropods, which lack a conspicuous shell, are commonly called slugs, and are scattered throughout groups that primarily include snails.
Carnivore	An animal that eats a diet consisting solely of meat is referred to as a carnivore.
Starfish	Starfish are marine invertebrates belonging to phylum Echinodermata, class Asteroidea. They exhibit a superficially radial symmetry, typically with five or more "arms" which radiate from an indistinct disk. They do not have movable skeletons, but instead possess a hydraulic water vascular system.
Crab	The term crab is often applied to several different groups of short (nose to tail) decapod crustaceans with thick exoskeletons, but only members of the Brachyura are true.
Species	Group of similarly constructed organisms capable of interbreeding and producing fertile offspring is a species.

Morphology	The scientific study of organic form, including both its development and function is morphology.
Adaptation	A biological adaptation is an anatomical structure, physiological process or behavioral trait of an organism that has evolved over a period of time by the process of natural selection such that it increases the expected long-term reproductive success of the organism.
Salinity	Salinity refers to the total amount of salts dissolved in seawater. It is generally expressed in parts per thousand.
Distribution	Distribution in pharmacology is a branch of pharmacokinetics describing reversible transfer of drug from one location to another within the body.
Clay	A soil constituent comprising particles smaller than 2 micrometers in diameter is clay. Clay is a generic term for an aggregate of hydrous silicate particles less than 2 ìm (micrometres) in diameter.
Erosion	Erosion is the displacement of solids (soil, mud, rock, and other particles) by the agents of wind, water, ice, movement in response to gravity, or living organisms.
Microorganism	A microorganism is an organism that is so small that it is microscopic (invisible to the naked eye). They are often illustrated using single-celled, or unicellular organisms; however, some unicellular protists are visible to the naked eye, and some multicellular species are microscopic.
Invertebrate	Invertebrate is a term coined by Jean-Baptiste Lamarck to describe any animal without a spinal column. It therefore includes all animals except vertebrates (fish, reptiles, amphibians, birds and mammals).
Bivalve	Bivalve refers to a member of a group of mollusks that includes clams, mussels, scallops, and oysters. They typically have two-part shells, with both parts being more or less symmetrical.
Mollusk	The mollusk is the large and diverse phylum Mollusca, which includes a variety of familiar creatures well-known for their decorative shells or as seafood. These range from tiny snails, clams, and abalone to the octopus, cuttlefish and squid (which are considered the most intelligent invertebrates).
Skeleton	In biology, the skeleton or skeletal system is the biological system providing physical support in living organisms.
Muscle contraction	A muscle contraction occurs when a muscle cell (called a muscle fiber) shortens. There are three general types: skeletal, heart, and smooth.
Muscle	Muscle is a contractile form of tissue. It is one of the four major tissue types, the other three being epithelium, connective tissue and nervous tissue. Muscle contraction is used to move parts of the body, as well as to move substances within the body.
Crustacean	Member of a group of marine arthropods that contains among others shrimps, crabs, crayfish, and lobsters is called crustacean.
Brachiopod	A brachiopod is an animala that superficially resembles clams and other bivalve mollusks, but the two halves of the brachiopod shell are dorsal and ventral to the animal rather than lateral, as in clams.
Mucus	Mucus is a slippery secretion of the lining of various membranes in the body (mucous membranes). Mucus aids in the protection of the lungs by trapping foreign particles that enter the nose during normal breathing. Additionally, it prevents tissues from drying out.
Tentacle	An elongate, extensible projection of the body of cnidarians and cephalopod mollusks that may be used for grasping, stinging, and immobilizing prey, and for locomotion is called tentacle.

Egg	An egg is the zygote, resulting from fertilization of the ovum. It nourishes and protects the embryo.
Dissolved oxygen	Oxygen gas molecules dissolved in water. Fish and other aquatic organisms are dependent on dissolved oxygen for respiration. Therefore concentration of dissolved oxygen is a measure of water quality.
Oxygen	Oxygen is a chemical element in the periodic table. It has the symbol O and atomic number 8. Oxygen is the second most common element on Earth, composing around 46% of the mass of Earth's crust and 28% of the mass of Earth as a whole, and is the third most common element in the universe.
Hydrogen sulfide	The gas that is produced in anoxic sediments is referred to as hydrogen sulfide.
Bacteria	The domain that contains procaryotic cells with primarily diacyl glycerol diesters in their membranes and with bacterial rRNA. Bacteria also is a general term for organisms that are composed of procaryotic cells and are not multicellular.
Zonation	The distribution of organisms in bands or regions corresponding to changes in ecological conditions along a continuum, for example, intertidal zonation and elevational zonation.
Redox potential	Redox potential is the tendency of a chemical species to acquire electrons and thereby be reduced. Each species has its own intrinsic reduction potential; the more positive the potential, the greater the species' affinity for electrons and tendency to be reduced.
Gastropod	A member of the largest group of mollusks, including snails and slugs is a gastropod.
Copepod	The group of small crustaceans that is an important member of marine and freshwater plankton communities is copepod. They are very important food organisms for small fish, whales, and other crustaceans such as krill in the ocean and in fresh water.
Diffusion	Diffusion refers to the spontaneous movement of particles of any kind from where they are more concentrated to where they are less concentrated.
Gradient	Gradient refers to a difference in concentration, pressure, or electrical charge between two regions.
Anaerobic organism	An anaerobic organism or anaerobe is any organism that does not require oxygen for growth. Obligate anaerobes will die when exposed to atmospheric levels of oxygen, while facultative anaerobes can use oxygen when it is present.
Aerobic organism	An aerobic organism is an organism that has an oxygen based metabolism. In a process known as cellular respiration, use oxygen to oxidize substrates (for example sugars and fats) in order to obtain energy.
Siphon	The tube-like extension through which water flows in and out of the mantle cavity in bivalves and cephalopods, and in tunicates is called siphon.
Metabolism	Metabolism is the biochemical modification of chemical compounds in living organisms and cells. This includes the biosynthesis of complex organic molecules (anabolism) and their breakdown (catabolism).
Anaerobic	An anaerobic organism is any organism that does not require oxygen for growth.
Anaerobic bacteria	Anaerobic bacteria refers to bacteria that do not need oxygen.
Heterotrophic	Incapable of synthesizing food and therefore dependent on other organisms for it is called heterotrophic.
Photosynthesis	Photosynthesis is a biochemical process in which plants, algae, and some bacteria harness the energy of light to produce food. Ultimately, nearly all living things depend on energy

produced from photosynthesis for their nourishment, making it vital to life on Earth.

Carbohydrate Carbohydrate is a chemical compound that contains oxygen, hydrogen, and carbon atoms. They consist of monosaccharide sugars of varying chain lengths and that have the general chemical formula $C_n(H_2O)_n$ or are derivatives of such.

Sugar A sugar is the simplest molecule that can be identified as a carbohydrate. These include monosaccharides and disaccharides, trisaccharides and the oligosaccharides. The term "glyco-" indicates the presence of a sugar in an otherwise non-carbohydrate substance.

Substrate A substrate is a molecule which is acted upon by an enzyme. Each enzyme recognizes only the specific substrate of the reaction it catalyzes. A surface in or on which an organism lives.

Hydrogen Hydrogen is a chemical element in the periodic table that has the symbol H and atomic number 1. At standard temperature and pressure it is a colorless, odorless, nonmetallic, univalent, tasteless, highly flammable diatomic gas.

Cyanobacteria Cyanobacteria are a phylum of bacteria that obtain their energy through photosynthesis. They are often referred to as blue-green algae, even though it is now known that they are not directly related to any of the other algal groups, which are all eukaryotes.

Decomposition Decomposition refers to the reduction of the body of a formerly living organism into simpler forms of matter.

Organic compound An organic compound is any member of a large class of chemical compounds whose molecules contain carbon, with the exception of carbides, carbonates, carbon oxides and gases containing carbon.

Fatty acid A fatty acid is a carboxylic acid (or organic acid), often with a long aliphatic tail (long chains), either saturated or unsaturated.

Alcohol Alcohol is a general term, applied to any organic compound in which a hydroxyl group (-OH) is bound to a carbon atom, which in turn is bound to other hydrogen and/or carbon atoms. The general formula for a simple acyclic alcohol is $C_nH_{2n+1}OH$.

Sulfur Sulfur is the chemical element in the periodic table that has the symbol S and atomic number 16. It is an abundant, tasteless, odorless, multivalent non-metal. Sulfur, in its native form, is a yellow crystaline solid. In nature, it can be found as the pure element or as sulfide and sulfate minerals.

Dominance Dominance is the state of having high social status relative to other individuals, who react submissively to dominant individuals.

Fermentation Fermentation is the anaerobic metabolic breakdown of a nutrient molecule, such as glucose, without net oxidation. Fermentation does not release all the available energy in a molecule; it merely allows glycolysis to continue by replenishing reduced coenzymes.

Sulfate reduction The process of sulfate use as an oxidizing agent, which results in the accumulation of reduced forms of sulfur such as sulfide, or incorporation of sulfur into organic molecules, usually as sulthydryl groups is called sulfate reduction.

Nutrition Nutrition refers to collectively, the processes involved in taking in, assimilating, and utilizing nutrients.

Sea cucumber The sea cucumber is an echinoderm of the class Holothuroidea, with an elongated body and leathery skin. Mostly found on the sea floor.

Residue A residue refers to a portion of a larger molecule, a specific monomer of a polysaccharide, protein or nucleic acid.

Amphipods Amphipods include over 7000 described species of small, shrimp-like crustaceans. Most

	amphipods are marine, although a few live in freshwater or are terrestrial. One species common to mid-Atlantic and estuarine waters is Jassa falcata.
Digestive tract	The digestive tract is the system of organs within multicellular animals which takes in food, digests it to extract energy and nutrients, and expels the remaining waste.
Browser	A type of herbivore whose nutrition generally comes from high growing plants, like trees, rather than a grazer that eats from the ground is a browser. Deer and goats are domesticated browsers, and elephants and giraffes are wild browsers.
Microbe	A microbe is an organism that is so small that it is microscopic (invisible to the naked eye). Microorganisms are often illustrated using single-celled, or unicellular organisms; however, some unicellular protists are visible to the naked eye, and some multicellular species are microscopic.
Diatom	Diatom refers to a unicellular photosynthetic alga with a unique, glassy cell wall containing silica.
Tissue	Group of similar cells which perform a common function is called tissue.
Complex carbohydrate	A complex carbohydrate consists of three or more monosaccharides bonded together; these are also known as polysaccharides. Complex carbohydrates are found in foods of plant origin.
Enzyme activity	Enzyme activity is the catalytic effect exerted by an enzyme.
Cellulose	A large polysaccharide composed of many glucose monomers linked into cable-like fibrils that provide structural support in plant cell walls is referred to as cellulose.
Cellulase	Cellulase refers to an enzyme that catalyzes the breakdown of the carbohydrate cellulose into its component glucose molecules; almost entirely restricted to microorganisms.
Fragmentation	Fragmentation is a form of asexual reproduction where an organism is split into fragments. The splitting may or may not be intentional. Each of these fragments develops into a mature, fully grown new organism, and if the organism is split any further the process is repeated.
Leaching	Leaching is the process of extracting a substance from a solid by dissolving it in a liquid. In the chemical processing industry, leaching is known as extraction.
Nitrogen	A colorless and tasteless and mostly inert diatomic non-metal gas that is an essential constituent of proteins is nitrogen.
Grazing	Grazing refers to consumption of a portion of a plant's tissues, generallyapplied to grasses and other herbaceous vegetation; removing plant or algal cover from a substrate, generally applied to aquatic systems.
Cell division	Cell division is the process by which a cell (called the parent cell) divides into two cells (called daughter cells). Cell division is usually a small segment of a larger cell cycle. In meiosis, however, a cell is permanently transformed and cannot divide again.
Crop	An organ, found in both earthworms and birds, in which ingested food is temporarily stored before being passed to the gizzard, where it is pulverized is the crop.
Cell	The cell is the structural and functional unit of all living organisms, and is sometimes called the "building block of life."
Detritus	Detritus is organic waste material from decomposing dead plants or animals.
Continental shelf	The shallow part of the seafloor immediately adjacent to a continent is called the continental shelf.
Phytoplankton	Phytoplankton refers to the autotrophic component of the plankton that drifts in the water column. Phytoplankton obtain energy through a process called photosynthesis, and so must live

	in the well-lit surface layer (termed the euphotic zone) of an ocean, sea, or lake.
Population	Group of organisms of the same species occupying a certain area and sharing a common gene pool is referred to as population.
Kelp	Kelp are large seaweeds, belonging to the brown algae and classified in the order Laminariales. Despite their appearance they are not grouped with the normal aquatic or land plants (kingdom Plantae), but instead are included in either kingdom Protista or Chromista.
Decomposer	Decomposer refers to an organism that derives its energy from organic wastes and dead organisms; also called a detritivore.
Micronutrient	Micronutrient refers to an element that an organism needs in very small amounts and that functions as a component or cofactor of enzymes.
Amino acid	An amino acid is any molecule that contains both amino and carboxylic acid functional groups. They are the basic structural building units of proteins. They form short polymer chains called peptides or polypeptides which in turn form structures called proteins.
Vitamin	A Vitamin is an organic molecule required by a living organism in minute amounts for proper health. An organism deprived of all sources of a particular vitamin will eventually suffer from disease symptoms specific to that vitamin.
Meiofauna	Meiofauna are small benthic invertebrates that live in both marine and fresh water environments. The term Meiofauna loosely defines a group of organisms by their size, larger than Microfauna but smaller than Macrofauna, rather than a taxonomic grouping.
Carbon	Carbon is a chemical element in the periodic table that has the symbol C and atomic number 6. An abundant nonmetallic, tetravalent element, carbon has several allotropic forms.
Frond	A frond is the leaf structure of a fern. The term is colloquially applied to the leaves of palms, cycads, and other plants with pinnately compound leaves. A significant difference is that, unlike the leaves of the latter, fern fronds bear the reproductive structures (spore-bearing structures) of the sporophyte plant.
Enzyme	An enzyme is a protein that catalyzes, or speeds up, a chemical reaction. They are essential to sustain life because most chemical reactions in biological cells would occur too slowly, or would lead to different products, without them.
Protein	A protein is a complex, high-molecular-weight organic compound that consists of amino acids joined by peptide bonds. They are essential to the structure and function of all living cells and viruses. Many are enzymes or subunits of enzymes.
Surfactant	A substance that decreases the surface tension of a liquid, allowing easier spreading, and the interfacial tension between two liquids is a surfactant.
Algae	The algae consist of several different groups of living organisms that capture light energy through photosynthesis, converting inorganic substances into simple sugars with the captured energy.
Renewable resource	A renewable resource is any natural resource that is depleted at a rate slower than the rate at which it regenerates.
Growth rate	A measure of the change in population size per individual per unit of time is called growth rate.
Standing stock	Standing stock refers to the total amount, or biomass, of an organism at a given time.
Fecal	Fecal is the solid excretory wastes of animals. Consists of undigested material passing through the gut and bacteria that have begun to feed on it.
Genus	In biology, a genus is a taxonomic grouping. That is, in the classification of living

organisms, a genus is considered to be distinct from other such genera. A genus has one or more species: if it has more than one species these are likely to be morphologically more similar than species belonging to different genera.

Ion — Ion refers to an atom or molecule that has gained or lost one or more electrons, thus acquiring an electrical charge.

Organ — Organ refers to a structure consisting of several tissues adapted as a group to perform specific functions.

Gill — An extension of the body surface of an aquatic animal, specialized for gas exchange and/or suspension feeding is called a gill.

Cellular metabolism — The sum of all chemical changes that take place in a cell through which energy and basic components are provided for essential processes, including the synthesis of new molecules and the breakdown and removal of others is called cellular metabolism.

Oxidation — Oxidation refers to the loss of electrons from a substance involved in a redox reaction; always accompanies reduction.

Sulfide — One of the minerals that is abundant in the hot water that seeps through hydrothermal vents is sulfide.

Hydrocarbon — A chemical compound composed only of the elements carbon and hydrogen is called hydrocarbon.

Surface layer — The upper layer of water that is mixed by wind, waves, and currents is the surface layer .

Natural selection — Natural selection is the process by which biological individuals that are endowed with favorable or deleterious traits end up reproducing more or less than other individuals that do not possess such traits.

Digestion — Digestion refers to the mechanical and chemical breakdown of food into molecules small enough for the body to absorb; the second main stage of food processing, following ingestion.

Plankton — Plankton are drifting organisms that inhabit the water column of oceans, seas, and bodies of fresh water. Plankton abundance and distribution are strongly dependent on factors such as ambient nutrients concentrations, the physical state of the water column, and the abundance of other plankton.

Cilia — Numerous short, hairlike structures projecting from the cell surface that enable locomotion are cilia.

Tide — The periodic, rhythmic rise and fall of the sea surface caused by changes in gravitational forces external to the Earth is referred to as the tide.

Barnacle — A barnacle is a type of arthropod belonging to infraclass Cirripedia in the subphylum Crustacea and is hence distantly related to crabs and lobsters. They live attached to surfaces and are typically enclosed by heavy calcareous plates are referred to as barnacles.

Sea urchin — A sea urchin is a spiny sea creature of the class Echinoidea found in oceans all over the world. Their shell, which biologists call the test, is globular in shape, and covered with spines.

Sessile — In botany, sessile means without a stalk, as in flowers or leaves that grow directly from the stem. In zoology, sessile organisms are those which are not able to move about. They are usually permanently attached to a solid substrate of some kind, such as a rock, or the hull of a ship in the case of barnacles.

Colony — Colony refers to a cluster or assemblage of microorganisms growing on the surface of an agar culture medium. A colony also refers to several individual organisms of the same species living closely together, usually for mutual benefit, such as stronger defences, the ability

	to attack bigger prey, etc.
Brittle star	A brittle star is an echinoderm with five flexible arms that radiate from a conspicuous central disk and tube feet that are used in feeding.
Tube feet	Tube feet are the many small tubular projections found most famously on the ventral face of a sea star's arms, but are characteristic of the water vascular system of the echinoderm phylum which also includes sea urchins, sand dollars and sea cucumbers and many other sea creatures.
Bolus	A bolus is any kind of ball-shaped organic structure of an organism or of its discharged substances..
Trait	In biology, a trait or character is a genetically inherited feature of an organism: Eye color is a character; brown eyes and blue eyes is a trait.
Downstream	Downstream refers to a relative position in DNA or RNA. Relative to the position on the strand, downstream is the region towards the 3' end of the strand.
Stipe	A stipe is a supportive structure that may be stemlike, as in seaweed, or a true leaf stem, as in ferns.
Cell wall	Cell wall refers to a protective layer external to the plasma membrane in plant cells, bacteria, fungi, and some protists; protects the cell and helps maintain its shape.
Sea squirts	Tunicates with a sac-like, attached body called a tunic as adults are referred to as sea squirts.
Stalk	The stalk is the longish piece that supports the seed-carrying parts of a plant, or, more simply, the stem. A leaf's petiole; the slender stem that supports the blade of a leaf and attaches it to a larger stem of the plant.
Upstream	Upstream refers to movement opposite the direction RNA follows when moving along a gene.
Fiber	Fiber is a class of materials that are continuous filaments or are in discrete elongated pieces, similar to lengths of thread. They are of great importance in the biology of both plants and animals, for holding tissues together.
Radius	The radius is the bone of the forearm that extends from the inside of the elbow to the thumb side of the wrist. The radius is situated on the lateral side of the ulna, which exceeds it in length and size.
Gorgonians	Colonial anthozoans that secrete a skeleton made of protein are referred to as gorgonians. They are horny corals, an example is a Sea Fan.
Polyp	Small, abnormal growth that arises from the epithelial lining is referred to as polyp. In zoology, a polyp is one of two forms of individuals found in many species of cnidarians. They are approximately cylindrical, elongated on the axis of the body.
Sponge	An invertebrates that consist of a complex aggregation of cells, including collar cells, and has a skeleton of fibers and/or spicules is a sponge. They are primitive, sessile, mostly marine, waterdwelling filter feeders that pump water through their matrix to filter out particulates of food matter.
Zooplankton	Nonphotosynthetic protists that feed on other plankton and are abundant in marine and freshwater environments are zooplankton.
Annelid	Annelid refers to member of a phylum of invertebrates that contains segmented worms, such as the earthworm and the clam worm.
Threshold	Electrical potential level at which an action potential or nerve impulse is produced is called threshold.

Biology	Biology is the branch of science dealing with the study of life. It is concerned with the characteristics, classification, and behaviors of organisms, how species come into existence, and the interactions they have with each other and with the environment.
Marine snow	In the deep ocean, marine snow is a continuous shower of mostly organic detritus falling from the upper layers of the water column. Its origin lies in activities within the productive photic zone.
Exudate	An exudate is any fluid that filters from the circulatory system into leisions or areas of inflamation. Its composition varies but generally includes water and the disolved solutes of the blood, some or all plasma proteins, white blood cells, platelets and red blood cells.
Silt	Silt refers to soil particles between the size of sand particles and clay particles, namely particles 0.002-0.2 mm in diameter.
Mantle cavity	A mantle skirt is a double fold of mantle that encloses a water space. This space is called the mantle cavity, and it is a central feature of mollusk biology, containing the mollusk's gills, anus, osphradium, nephridiopores, and gonopores.
Filament	The stamen is the male organ of a flower. Each stamen generally has a stalk called the filament, and, on top of the filament, an anther. The filament is a long chain of proteins, such as those found in hair, muscle, or in flagella.
Ventral	The surface or side of the body normally oriented upwards, away from the pull of gravity, is the dorsal side; the opposite side, typically the one closest to the ground when walking on all legs, swimming or flying, is the ventral side.
Relative abundance	Differences in the abundance of species within a community is relative abundance.
Insertion	At DNA level, an insertion means the insertion of a few base pairs into a genetic sequence. This can often happen in microsatellite regions due to the DNA polymerase slipping.
Lens	The lens or crystalline lens is a transparent, biconvex structure in the eye that, along with the cornea, helps to refract light to focus on the retina. Its function is thus similar to a man-made optical lens.
Photosynthetic pigment	A photosynthetic pigment is a pigment that is present in chloroplasts or photosynthetic bacteria and captures the light energy necessary for photosynthesis.
Dorsal	In anatomy, the dorsal is the side in which the backbone is located. This is usually the top of an animal, although in humans it refers to the back.
Refraction	Refraction is the change in direction of a wave due to a change in the velocity of the wave. It's most often demonstrated in optics, but may happen for any wave that interacts with a medium, such as sound waves.
Septum	A septum, in general, is a wall separating two cavities or two spaces containing a less dense material. The muscle wall that divides the heart chambers.
Cephalopod	Cephalopod refers to a member of a group of mollusks that includes squids and octopuses, characterized by bilateral body symmetry.
Predation	Interaction in which one organism uses another, called the prey, as a food source is referred to as predation.
Predator	A predator is an animal or other organism that hunts and kills other organisms for food in an act called predation.
Olfaction	Olfaction, the sense of odor, is the detection of chemicals dissolved in air.In vertebrates smells are sensed by the olfactory epithelium located in the nasal cavity and first processed

	by the olfactory bulb in the olfactory system. In insects smells are sensed by sensilia located on the antenna and first processed by the antennal lobe.
Cone	Cone refers to a reproductive structure of gymnosperms that produces pollen in males or eggs in females.
Virulence	The degree or intensity of pathogenicity of an organism as indicated by case fatality rates and/or ability to invade host tissues and cause disease is referred to as the virulence.
Exoskeleton	An exoskeleton is an external anatomical feature that supports and protects an animal's body. Many invertebrate animals such as insects, crustaceans and shellfish have an exoskeleton.
Radula	A toothed, grasping organ found in many mollusks, used to scrape up or shred food is referred to as a radula.
Secretion	Secretion is the process of segregating, elaborating, and releasing chemicals from a cell, or a secreted chemical substance or amount of substance.
Macrophage	Macrophage is a cell found in tissues that are responsible for phagocytosis of pathogens, dead cells and cellular debris. They are part of the innate immune system. Their main role is the removal of pathogens and necrotic debris.
Esophagus	The esophagus, or gullet is the muscular tube in vertebrates through which ingested food passes from the mouth area to the stomach. Food is passed through the esophagus by using the process of peristalsis.
Sagittal	A sagittal plane is an X-Z plane, perpendicular to the ground and to the coronal plane, which separates left from right. The midsagittal plane is the specific sagittal plane that is exactly in the middle of the body.
Buccal cavity	The buccal cavity is the opening through which an animal or human takes in food and water.
Calcareous	Calcareous refers to a sediment, sedimentary rock, or soil type which is formed from or contains a high proportion of calcium carbonate in the form of calcite or aragonite. It is also used to refer to relatively alkaline soil.
Ligament	A ligament is a short band of tough fibrous connective tissue composed mainly of long, stringy collagen fibres. They connect bones to other bones to form a joint. (They do not connect muscles to bones.)
Isopod	Isopod is one of the most diverse orders of Crustaceans, with many species living in all environments, and are common in shallow marine waters. Unlike most crustaceans, isopods are successful on land (suborder Oniscidea, woodlice), although their greatest diversity remains in the deep sea (suborder Asellota).
Coral reef	A coral reef is a type of biotic reef that develops in tropical waters by the growth of coralline algae, hermatypic corals, and other marine organisms.
Green algae	Green algae refers to photosynthetic protists that include L.nicellular, colonial, and multicellular species with grass green chloroplasts; closely related to true plants.
Patella	The patella is a thick, triangular bone which articulates with the femur and covers and protects the front of the knee joint. The patella increases the leverage that the tendon can exert on the femur by increasing the angle at which it acts.
Flexor	A muscle or tendon which bends a limb, part of a limb, or part of the body is a flexor.
Extensor	A muscle or tendon that straightens or extends a limb is called an extensor.
Symbiotic nitrogen-fixing bacteria	Bacteria that live in the roots of certain kinds of plants, where they convert nitrogen gas molecules into compounds that plants can use is called symbiotic nitrogen-fixing bacteria.

Salt marsh	Salt marsh refers to a grassy area that extends along the shores of estuaries and sheltered coasts in temperate and subpolar regions.
Zostera	Zostera is a small genus of widely distributed seagrass, commonly called eelgrass. It contains twelve species. Zostera is found on sandy sunstrates or in esturaries submerged or partially floating. Most Zostera are perennial, they have long, bright green, ribbon-like leaves, about 1 cm wide.
Ruminant	A ruminant is any hooved animal that digests its food in two steps, first by eating the raw material and regurgitating a semi-digested form known as cud, then eating the cud They also have four stomach chambers.
Mammal	Homeothermic vertebrate characterized especially by the presence of hair and mammary glands is a mammal.
Protozoa	Protozoa are single-celled eukaryotes (organisms with nuclei) that show some characteristics usually associated with animals, most notably mobility and heterotrophy. They are often grouped in the kingdom Protista together with the plant-like algae.
Food chain	Food chain refers to a sequence of food transfers from one species to another within an ecosystem.
Ecosystem	In general terms an ecosystem can be thought of as an assemblage of organisms (plant, animal and other living organisms living together with their environment, functioning as a loose unit. That is, a dynamic and complex whole, interacting as an "ecological unit".
Detritivore	Detritivore refers to an organism that derives its energy from organic wastes and dead organisms; also called a decomposer.
Ecology	Ecology is the scientific study of the distribution and abundance of living organisms and how these properties are affected by interactions between the organisms and their environment.
Systematics	In biology, systematics is the study of the diversity of organism characteristics, and especially how they relate evolutionarily. The term "systematics" may or may not overlap with "taxonomy" which concerns itself with scientific classification of species and other taxa.
Biotic community	An association of plants, animals, and other organisms living together in a specific habitat is a biotic community.
Microhabitat	A microhabitat is the immediate surroundings and other physical factors of an individual plant or animal within its habitat.
Biomass	Biomass is organic non-fossil material, collectively. In other words, biomass comprises the mass of all biological organisms, dead or alive, excluding biological mass that has been transformed by geological processes into substances such as coal or petroleum.
Interstitial fauna	Animals living between sediment particle are interstitial fauna.
Sea anemone	Named after a terrestrial flower, the sea anemone is a group of water-dwelling, filter feeding animals of the order Actinaria. As a cnidarian, it is closely related to coral and jellyfish.
Boundary layer	Boundary layer refers to a layer of still or slow-moving water or air close to the surface of an object.
Limnology	Limnology is a discipline that concerns the study of inland waters (both saline and fresh), specifically lakes, ponds and rivers (both natural and manmade), including their biological, physical, chemical, and hydrological aspects.
Bryozoan	Colonial animals that build stony skeletons of calcium carbonate and superficially resemble

	coral is referred to as bryozoan.
Absorption	Absorption is a physical or chemical phenomenon or a process in which atoms, molecules, or ions enter some bulk phase - gas, liquid or solid material. In nutrition, amino acids are broken down through digestion, which begins in the stomach.
Filtration	Filtration involved in passive transport is the movement of water and solute molecules across the cell membrane due to hydrostatic pressure by the cardiovascular system.
In vivo	In vivo is used to indicate the presence of a whole/living organism, in distinction to a partial or dead organism, or a computer model. Animal testing and clinical trials are forms of in vivo research.
Sea stars	Echinoderms with five or more radiating arms and tube feet that are used in locomotion are sea stars.
Cancer	Cancer is a class of diseases or disorders characterized by uncontrolled division of cells and the ability of these cells to invade other tissues, either by direct growth into adjacent tissue through invasion or by implantation into distant sites by metastasis.
Crustacea	The animal phylum that includes mostly aquatic animals such as crabs, lobsters, crayfish, and shrimp is referred to as crustacea.
Decapoda	The decapoda are an order of crustaceans within the class Malacostraca, including many familiar groups, such as crabs, lobsters, prawns and shrimps, but also some families that are less well known.
Zoology	Zoology is the biological discipline which involves the study of animals.
Microbiology	Microbiology is the study of microorganisms, including viruses, viroids, prions, prokaryotes and simple eukaryotes (e.g. molds and fungi).
Host	Host is an organism that harbors a parasite, mutual partner, or commensal partner; or a cell infected by a virus.
Calcification	A process in which tissue becomes hardened due to deposits of calcium salts. Calcification of blood vessels plays a role in the development of atherosclerosis.
Temperate	In virology, temperate refers to the life cycle a phage is able to perform. The phage usually integrates its genome into its host chromosome, becoming a prophage.
Functional group	Functional group is a submolecular structural motif, characterized by specific elemental composition and connectivity, that confer reactivity upon the molecule that contains it.

Mangrove	Mangrove are woody trees or shrubs that grow in coastal habitats or mangal (Hogarth, 1999), for which the term mangrove swamp also would apply.
Habitat	Habitat refers to a place where an organism lives; an environmental situation in which an organism lives.
Crustacean	Member of a group of marine arthropods that contains among others shrimps, crabs, crayfish, and lobsters is called crustacean.
Species	Group of similarly constructed organisms capable of interbreeding and producing fertile offspring is a species.
Population	Group of organisms of the same species occupying a certain area and sharing a common gene pool is referred to as population.
Predation	Interaction in which one organism uses another, called the prey, as a food source is referred to as predation.
Predator	A predator is an animal or other organism that hunts and kills other organisms for food in an act called predation.
Interspecific competition	Competition between populations of two or more species for one or more of the same limited resources of food, sunlight, water, soil, nutrients, or space. Interspecific competition may inhibit population growth and help structure communities.
Biological interaction	A biological interaction results from the fact that organisms in an ecosystem interact with each other, in the natural world, no organism is an autonomous entity isolated from its surroundings.
Dominant species	Dominant species refers to those species in a community that have the highest abundance or highest biomass. These species exert a powerful control over the occurrence and distribution of other species.
Intertidal zone	The part of a beach that lies between average high tide and average low tide is called the intertidal zone.
Zonation	The distribution of organisms in bands or regions corresponding to changes in ecological conditions along a continuum, for example, intertidal zonation and elevational zonation.
Tide	The periodic, rhythmic rise and fall of the sea surface caused by changes in gravitational forces external to the Earth is referred to as the tide.
Temperate	In virology, temperate refers to the life cycle a phage is able to perform. The phage usually integrates its genome into its host chromosome, becoming a prophage.
Boreal	Boreal is usually applied to ecosystems localized to subarctic (Northern hemisphere) and subantarctic (Southern hemisphere) zones, although Austral is also used for the latter. A boreal forest, also known as the taiga, is the set of forest ecosystems that can survive in northern, specifically subarctic, regions.
Gastropod	A member of the largest group of mollusks, including snails and slugs is a gastropod.
Lichen	Lichen is a symbiotic organism in a mutualistic association between a fungus and an alga or between a fungus and a cyanobacterium.
Biotic	Biotic refers to life or living organisms.
Snail	The name snail applies to most members of the molluscan class Gastropoda that have coiled shells. Other gastropods, which lack a conspicuous shell, are commonly called slugs, and are scattered throughout groups that primarily include snails.
Barnacle	A barnacle is a type of arthropod belonging to infraclass Cirripedia in the subphylum

	Crustacea and is hence distantly related to crabs and lobsters. They live attached to surfaces and are typically enclosed by heavy calcareous plates are referred to as barnacles.
Mussel	A mussel is a bivalve mollusk that can be found in lakes, rivers, creeks, intertidal areas, and throughout the ocean. The saltwater mussels (family Mytilidae) and freshwater mussels (family Unionidae) are not thought to be closely related and are grouped in different subclasses, despite considerable similarities in appearance.
Mudflat	Mudflat is a coastal wetland that forms when mud is deposited by the tides or rivers, they are found in sheltered areas such as bays, bayous, lagoons, and estuaries.
Dominance	Dominance is the state of having high social status relative to other individuals, who react submissively to dominant individuals.
Oxygen	Oxygen is a chemical element in the periodic table. It has the symbol O and atomic number 8. Oxygen is the second most common element on Earth, composing around 46% of the mass of Earth's crust and 28% of the mass of Earth as a whole, and is the third most common element in the universe.
Sediment	Sediment is any particulate matter that can be transported by fluid flow and which eventually is deposited as a layer of solid particles on the bed or bottom of a body of water or other liquid.
Salinity	Salinity refers to the total amount of salts dissolved in seawater. It is generally expressed in parts per thousand.
Invertebrate	Invertebrate is a term coined by Jean-Baptiste Lamarck to describe any animal without a spinal column. It therefore includes all animals except vertebrates (fish, reptiles, amphibians, birds and mammals).
Adaptation	A biological adaptation is an anatomical structure, physiological process or behavioral trait of an organism that has evolved over a period of time by the process of natural selection such that it increases the expected long-term reproductive success of the organism.
Protein	A protein is a complex, high-molecular-weight organic compound that consists of amino acids joined by peptide bonds. They are essential to the structure and function of all living cells and viruses. Many are enzymes or subunits of enzymes.
Seaweed	Phycologists consider seaweed to refer any of a large number of marine benthic algae that are multicellular, macrothallic (large-bodied), and thus differentiated from most algae that tend to be microscopic in size.
Sessile	In botany, sessile means without a stalk, as in flowers or leaves that grow directly from the stem. In zoology, sessile organisms are those which are not able to move about. They are usually permanently attached to a solid substrate of some kind, such as a rock, or the hull of a ship in the case of barnacles.
Microhabitat	A microhabitat is the immediate surroundings and other physical factors of an individual plant or animal within its habitat.
Gradient	Gradient refers to a difference in concentration, pressure, or electrical charge between two regions.
Evaporative cooling	Evaporative cooling is a system in which latent heat of evaporation is used to carry heat away from an object to cool it. The latent heat contains a considerable amount of energy, and carries away more heat than if the same temperature liquid was simply removed physically.
Subtidal zone	The subtidal zone is the area below the range of tidal exposure; always covered by water.
Sunlight	Sunlight in the broad sense is the total spectrum of electromagnetic radiation given off by the Sun.

Radiation	The emission of electromagnetic waves by all objects warmer than absolute zero is referred to as radiation.
Amino acid	An amino acid is any molecule that contains both amino and carboxylic acid functional groups. They are the basic structural building units of proteins. They form short polymer chains called peptides or polypeptides which in turn form structures called proteins.
Cell	The cell is the structural and functional unit of all living organisms, and is sometimes called the "building block of life."
Genus	In biology, a genus is a taxonomic grouping. That is, in the classification of living organisms, a genus is considered to be distinct from other such genera. A genus has one or more species: if it has more than one species these are likely to be morphologically more similar than species belonging to different genera.
Bivalve	Bivalve refers to a member of a group of mollusks that includes clams, mussels, scallops, and oysters. They typically have two-part shells, with both parts being more or less symmetrical.
Carnivore	An animal that eats a diet consisting solely of meat is referred to as a carnivore.
Starfish	Starfish are marine invertebrates belonging to phylum Echinodermata, class Asteroidea. They exhibit a superficially radial symmetry, typically with five or more "arms" which radiate from an indistinct disk. They do not have movable skeletons, but instead possess a hydraulic water vascular system.
Limpet	Limpet is a marine mollusk in the family Acmaeidae with flattened, cone-shaped shells. They live throughout the intertidal zone, from the high zone (upper littoral) to the shallow subtidal on the rocky coasts of most oceans. They can be commonly found attached to rocks, looking like little disks or bumps on the rock surface.
Suspension feeder	An animal that extracts food particles suspended in the surrounding water for food is referred to as a suspension feeder.
Phytoplankton	Phytoplankton refers to the autotrophic component of the plankton that drifts in the water column. Phytoplankton obtain energy through a process called photosynthesis, and so must live in the well-lit surface layer (termed the euphotic zone) of an ocean, sea, or lake.
Metabolic waste	Metabolic waste is the substances left over from metabolic processes and cannot be used by other processes, and must therefore be excreted.
Metabolic rate	Energy expended by the body per unit time is called metabolic rate.
Hemoglobin	Hemoglobin is the iron-containing oxygen-transport metalloprotein in the red cells of the blood in mammals and other animals. Hemoglobin transports oxygen from the lungs to the rest of the body, such as to the muscles, where it releases the oxygen load.
Pigment	Pigment is any material resulting in color in plant or animal cells which is the result of selective absorption.
Blood	Blood is a circulating tissue composed of fluid plasma and cells. The main function of blood is to supply nutrients (oxygen, glucose) and constitutional elements to tissues and to remove waste products.
Gill	An extension of the body surface of an aquatic animal, specialized for gas exchange and/or suspension feeding is called a gill.
Salt marsh	Salt marsh refers to a grassy area that extends along the shores of estuaries and sheltered coasts in temperate and subpolar regions.
Crab	The term crab is often applied to several different groups of short (nose to tail) decapod crustaceans with thick exoskeletons, but only members of the Brachyura are true.

Algae	The algae consist of several different groups of living organisms that capture light energy through photosynthesis, converting inorganic substances into simple sugars with the captured energy.
Phototactic	Phototactic describes the movement of a cell (eg a gamete) or a motile organism towards or away from a light source.
Desert	A desert is a landscape form or region that receives little precipitation - less than 250 mm (10 in) per year. It is a biome characterized by organisms adapted to sparse rainfall and rapid evaporation.
Mole	The atomic weight of a substance, expressed in grams. One mole is defined as the mass of 6.0222 3 1023 atoms.
Antennae	Antennae are paired appendages connected to the front-most segments of arthropods. In crustaceans, they are biramous and present on the first two segments of the head, with the smaller pair known as antennules. In insects, olfactory receptors on the antennae bind to odor molecules, including pheromones.
Migration	Migration occurs when living things move from one biome to another. In most cases organisms migrate to avoid local shortages of food, usually caused by winter. Animals may also migrate to a certain location to breed, as is the case with some fish.
Distribution	Distribution in pharmacology is a branch of pharmacokinetics describing reversible transfer of drug from one location to another within the body.
Morphology	The scientific study of organic form, including both its development and function is morphology.
Bladder	A hollow muscular storage organ for storing urine is a bladder.
Byssal thread	Byssal thread denotes strong threads secreted by mussels to attach to rocks and large, generally heavy objects in the intertidal zone. They range to 6 centimeters in length.
Holdfast	A rootlike structure that anchors a seaweed is called holdfast.
Biota	Biota is the plant and animal life of a region or area or geological period. The biota of the Earth is known as the biosphere.
Integument	Integument refers to the natural outer covering layers of an animal. Develops from the ectoderm.
Limiting factor	In biology, agricultural science, physiology, and ecology, a limiting factor is one that controls a process, such as organism growth or species population size or distribution. The availability of food, predation pressure, or availability of shelter are examples of factors that could be limiting for a species population in a specific area.
Abiotic	Abiotic refers to the nonliving portion of an ecosystem. It includes soil, rock, water, and the atmosphere.
Erosion	Erosion is the displacement of solids (soil, mud, rock, and other particles) by the agents of wind, water, ice, movement in response to gravity, or living organisms.
Whelk	A whelk is a large marine gastropod (snail) found in temperate waters. Whelks are different from, but sometimes confused as, conches (pronounced "Konks"). They are scavengers and carnivores, equipped with an extensible proboscis that is tipped with a file-like radula.
Intraspecific competition	Competition between individuals of the same species in a community for a limited resource is referred to as intraspecific competition.
Dispersal	Biological dispersal refers to those processes by which a species maintains or expands the distribution of a population. Dispersal implies movement—movement away from an existing

	population or away from the parent organisms.
Biology	Biology is the branch of science dealing with the study of life. It is concerned with the characteristics, classification, and behaviors of organisms, how species come into existence, and the interactions they have with each other and with the environment.
Ecology	Ecology is the scientific study of the distribution and abundance of living organisms and how these properties are affected by interactions between the organisms and their environment.
Kelp	Kelp are large seaweeds, belonging to the brown algae and classified in the order Laminariales. Despite their appearance they are not grouped with the normal aquatic or land plants (kingdom Plantae), but instead are included in either kingdom Protista or Chromista.
Barrier reef	A type of coral reef that develops at some distance from the coast creating a lagoon between itself and the coast is called a barrier reef.
Coral	Group of cnidarians having a calcium carbonate skeleton that participate in the formation of reefs is coral.
Colonization	Colonization is the act where life forms move into a distant area where their kind is sparse or not yet existing at all and set up new settlements in the area. Colonization applies to all life forms in a sense though it is most often used in reference to insects and humans.
Coral reef	A coral reef is a type of biotic reef that develops in tropical waters by the growth of coralline algae, hermatypic corals, and other marine organisms.
Zooplankton	Nonphotosynthetic protists that feed on other plankton and are abundant in marine and freshwater environments are zooplankton.
Surface water	Includes all bodies of water, lakes, rivers, ponds, and so on that are on the surface of the earth, in contrast to groundwater, which lies below the surface is surface water.
Egg	An egg is the zygote, resulting from fertilization of the ovum. It nourishes and protects the embryo.
Stratification	Stratification is the process of pretreating seeds to simulate natural conditions that a seed must endure before germination. Many seed species have what is called an embryonic dormancy and generally speaking will not sprout until this dormancy is broken.
Lagoon	A lagoon is a body of comparatively shallow salt water separated from the deeper sea by a shallow or exposed sandbank, coral reef, or similar feature. Thus, the enclosed body of water behind a barrier reef or barrier islands or enclosed by an atoll reef is called a lagoon.
Siphon	The tube-like extension through which water flows in and out of the mantle cavity in bivalves and cephalopods, and in tunicates is called siphon.
Aromatic	In chemistry, an aromatic molecule is one in which a conjugated ring of unsaturated bonds, lone pairs, or empty orbitals exhibit a stabilization stronger than would be expected by the stabilization of conjugation alone.
Mollusk	The mollusk is the large and diverse phylum Mollusca, which includes a variety of familiar creatures well-known for their decorative shells or as seafood. These range from tiny snails, clams, and abalone to the octopus, cuttlefish and squid (which are considered the most intelligent invertebrates).
Character displacement	Divergence in the characteristics of two otherwise similar species where their ranges overlap, caused by the selective effects of competition between the species in the area of overlap is a character displacement.
Genotype	The genotype is the specific genetic makeup (the specific genome) of an individual, usually in the form of DNA. It codes for the phenotype of that individual.

Diatom	Diatom refers to a unicellular photosynthetic alga with a unique, glassy cell wall containing silica.
Evolution	In biology, evolution is the process by which novel traits arise in populations and are passed on from generation to generation. Its action over large stretches of time explains the origin of new species and ultimately the vast diversity of the biological world.
Genetics	Genetics is the science of genes, heredity, and the variation of organisms.
Annuals	Plants that grow from seed, flower, set seed, and die, thus completing their life cycle in a single year are called annuals.
Moss	Moss is a small plant rarely taller than 2 inches (50 mm). They typically grow close together in clumps or mats in damp or shady locations. They do not have flowers and their simple leaves cover the thin wiry stems. At certain times they produces spore capsules which may appear as beak-like capsules borne aloft on thin stalks.
Keystone species	Species that are not usually abundant in a community yet exert strong control on community structure by the nature of their ecological roles or niches are keystone species.
Clear-cutting	Clear-cutting refers to the practice of cutting all trees in an area, regardless of species, size, quality, or age.
Herbivore	A herbivore is an animal that is adapted to eat primarily plant matter
Coralline algae	Red algae that deposit calcium carbonate in their thallus are called the coralline algae.
Barren	An area with sparse vegetation owing to some physical or chemical property of the soil is referred to as barren.
Grazing	Grazing refers to consumption of a portion of a plant's tissues, generallyapplied to grasses and other herbaceous vegetation; removing plant or algal cover from a substrate, generally applied to aquatic systems.
Chiton	A mollusc that has a shell divided into eight overlapping plates is a chiton.
Morula	A morula is an embryo at an early stage of embryonic development, consisting of approximately 4-16 cells (called blastomeres). The morula is produced by embryonic cleavage, the rapid cell division of the zygote.
Canopy	Canopy refers to the leaf-bearing part of a tree. Collectively the aggregate of the leaves and branches of the larger woody plants of an ecological community.
Horseshoe crab	Horseshoe crab refers to a bottom-dwelling arthropod that is more closely related to spiders than crabs, that belongs to the phylum Arthropoda.
Benthos	Benthos are the organisms and habitats of the sea floor; in freshwater biology they are the organisms and habitats of the bottoms of lakes, rivers, and creeks.
Crustacea	The animal phylum that includes mostly aquatic animals such as crabs, lobsters, crayfish, and shrimp is referred to as crustacea.
Annelid	Annelid refers to member of a phylum of invertebrates that contains segmented worms, such as the earthworm and the clam worm.
Food web	A network of interconnecting food chains is referred to as a food web.
Frond	A frond is the leaf structure of a fern. The term is colloquially applied to the leaves of palms, cycads, and other plants with pinnately compound leaves. A significant difference is that, unlike the leaves of the latter, fern fronds bear the reproductive structures (spore-bearing structures) of the sporophyte plant.
Spore	Spore refers to a differentiated, specialized form that can be used for dissemination, for

	survival of adverse conditions because of its heat and dessication resistance, and/or for reproduction. They are usually unicellular and may develop into vegetative organisms or gametes. A spore may be produced asexually or sexually and are of many types.
Microorganism	A microorganism is an organism that is so small that it is microscopic (invisible to the naked eye). They are often illustrated using single-celled, or unicellular organisms; however, some unicellular protists are visible to the naked eye, and some multicellular species are microscopic.
Bacteria	The domain that contains procaryotic cells with primarily diacyl glycerol diesters in their membranes and with bacterial rRNA. Bacteria also is a general term for organisms that are composed of procaryotic cells and are not multicellular.
Stipe	A stipe is a supportive structure that may be stemlike, as in seaweed, or a true leaf stem, as in ferns.
Growth rate	A measure of the change in population size per individual per unit of time is called growth rate.
Tissue	Group of similar cells which perform a common function is called tissue.
Community ecology	The study of how interactions between species affect community structure and organization is referred to as community ecology.
Threshold	Electrical potential level at which an action potential or nerve impulse is produced is called threshold.
Population density	The number of organisms of a species per unit area is called population density.
Deposit feeder	A deposit feeder is an organism that feeds on the particles of matter in the soil, usually the top sediment where it is filled with organic matter.
Water column	The vertical column of seawater that extends from the surface to the bottom is called water column.
Renewable resource	A renewable resource is any natural resource that is depleted at a rate slower than the rate at which it regenerates.
Oligochaete	The Oligochaete or "few-bristled" worms are well-segmented Annelids, most with a spacious coelom that is used as a hydroskeleton. Their setae (chaetae) or "bristles" are generally few in number and they lack the parapodia of the polychaeta.
Budding	A means of asexual reproduction whereby a new individual developed from an outgrowth of a parent splits off and lives independently is referred to as budding.
Absorption	Absorption is a physical or chemical phenomenon or a process in which atoms, molecules, or ions enter some bulk phase - gas, liquid or solid material. In nutrition, amino acids are broken down through digestion, which begins in the stomach.
Primary productivity	Primary productivity is the amount of light energy converted to chemical energy in a given amount of time. It usually corresponds to the amount of photosynthesis that occurred.
Downstream	Downstream refers to a relative position in DNA or RNA. Relative to the position on the strand, downstream is the region towards the 3' end of the strand.
Population cycle	A population cycle in zoology is a phenomenon where populations rise and fall over a predictable period of time, because there are a number of factors which influence population change such as availability of food, predators, diseases and climate.
Carbon	Carbon is a chemical element in the periodic table that has the symbol C and atomic number 6. An abundant nonmetallic, tetravalent element, carbon has several allotropic forms.

Pulse	The rhythmic stretching of the arteries caused by the pressure of blood forced through the arteries by contractions of the ventricles during systole is a pulse.
Open system	Open system refers to any system that exchanges chemicals and energy with its surroundings. All organisms are open systems.
Watershed	The total land area that drains directly or indirectly into a particular stream or river. The watershed is generally named for the stream or river into which it drains.
Estuary	An estuary is a semi-enclosed coastal body of water which has a free connection with the open sea and within which sea water mixes with fresh water. The key feature of an estuary is that it is a mixing place for sea water and a stream or river to supply fresh water.
Pruning	In gardening, pruning is the practice of removing diseased, overmature, or otherwise unwanted portions from a woody plant.
Overfishing	Overfishing is a situation where one or more fish stocks are reduced below predefined levels of acceptance by fishing activities. Biological overfishing occurs when fishing mortality has reached a level where the stock biomass has negative marginal growth .
Pollution	Any environmental change that adversely affects the lives and health of living things is referred to as pollution.
Insect	An arthropod that usually has three body segments , three pairs of legs, and one or two pairs of wings is called an insect. They are the largest and (on land) most widely-distributed taxon within the phylum Arthropoda. They comprise the most diverse group of animals on the earth, with around 925,000 species described
Upstream	Upstream refers to movement opposite the direction RNA follows when moving along a gene.
Sea urchin	A sea urchin is a spiny sea creature of the class Echinoidea found in oceans all over the world. Their shell, which biologists call the test, is globular in shape, and covered with spines.
Echinoderm	Echinoderm is a phylum of marine animals found in the ocean at all depths. This phylum dates back to the lower Cambrian period and represents about 7000 living species and 13000 extinct ones. 6 classes made it to the Modern Era.
Hyperosmotic	The condition in which a solution has a higher osmotic concentration than that of a second solution is called hyperosmotic.
Tissue fluid	Tissue fluid is a solution which bathes and surrounds the cells of multicellular animals and is also called the intercellular fluid.
Species richness	Species richness is perhaps the simplest measure of biodiversity. The larger the number, the more species in an area. It is referred to in equations as S. There is a strong inverse correlation in many groups between species richness and latitude - the further from the equator you go, the fewer species you will find.
Salt	Salt is a term used for ionic compounds composed of positively charged cations and negatively charged anions, so that the product is neutral and without a net charge.
Rhizome	A rhizome is a usually-underground, horizontal stem of a plant that often sends out roots and shoots from its nodes.
Root	In vascular plants, the root is that organ of a plant body that typically lies below the surface of the soil. However, this is not always the case, since a root can also be aerial (that is, growing above the ground) or aerating (that is, growing up above the ground or especially above water).
Stem	Stem refers to that part of a plant's shoot system that supports the leaves and reproductive

	structures.
Leaf	In botany, a leaf is an above-ground plant organ specialized for photosynthesis. For this purpose, a leaf is typically flat (laminar) and thin, to expose the chloroplast containing cells (chlorenchyma tissue) to light over a broad area, and to allow light to penetrate fully into the tissues.
Soil	Soil is material capable of supporting plant life. Soil forms through a variety of soil formation processes, and includes weathered rock "parent material" combined with dead and living organic matter and air.
Shoot	In botany, the shoot is one of two primary sections of a plant; the other is the root. The shoot refers to what is generally the upper portion of a plant, and consists of stems, leaves, flowers, and fruits. It is derived from the embryonic epicotyl, the portion of the embryo above the point of attachment to the seed leaves (cotyledons).
Flower	A flower is the reproductive structure of a flowering plant. The flower structure contains the plant's reproductive organs, and its function is to produce seeds through sexual reproduction.
Enzyme	An enzyme is a protein that catalyzes, or speeds up, a chemical reaction. They are essential to sustain life because most chemical reactions in biological cells would occur too slowly, or would lead to different products, without them.
Clone	A group of genetically identical cells or organisms derived by asexual reproduction from a single parent is called a clone.
Mammal	Homeothermic vertebrate characterized especially by the presence of hair and mammary glands is a mammal.
Cellulose	A large polysaccharide composed of many glucose monomers linked into cable-like fibrils that provide structural support in plant cell walls is referred to as cellulose.
Germination	The resumption of growth and development by a spore or seed is called germination. It is the process in botany where growth emerges from a resting stage.
Fecal	Fecal is the solid excretory wastes of animals. Consists of undigested material passing through the gut and bacteria that have begun to feed on it.
Fauna	Fauna (animals), a collective term for animal life, as distinct from flora (plant life).
Decomposition	Decomposition refers to the reduction of the body of a formerly living organism into simpler forms of matter.
Fragmentation	Fragmentation is a form of asexual reproduction where an organism is split into fragments. The splitting may or may not be intentional. Each of these fragments develops into a mature, fully grown new organism, and if the organism is split any further the process is repeated.
Detritivore	Detritivore refers to an organism that derives its energy from organic wastes and dead organisms; also called a decomposer.
Substrate	A substrate is a molecule which is acted upon by an enzyme. Each enzyme recognizes only the specific substrate of the reaction it catalyzes. A surface in or on which an organism lives.
Microbe	A microbe is an organism that is so small that it is microscopic (invisible to the naked eye). Microorganisms are often illustrated using single-celled, or unicellular organisms; however, some unicellular protists are visible to the naked eye, and some multicellular species are microscopic.
Particulate matter	In air pollution, solid particles and liquid droplets, as opposed to material uniformly dispersed among air molecules is particulate matter.

Somatic	The term somatic refers to the body. It also refers to the part of the nervous system that controls voluntary movement and senzation and judges relative effort and weight, called proprioception.
Food chain	Food chain refers to a sequence of food transfers from one species to another within an ecosystem.
Nutrition	Nutrition refers to collectively, the processes involved in taking in, assimilating, and utilizing nutrients.
Detritus	Detritus is organic waste material from decomposing dead plants or animals.
Nitrogen	A colorless and tasteless and mostly inert diatomic non-metal gas that is an essential constituent of proteins is nitrogen.
Chromosome	A chromosome is, minimally, a very long, continuous piece of DNA, which contains many genes, regulatory elements and other intervening nucleotide sequences.
Spawning	A method of external fertilization in which male and female parents shed gametes into the water, and sperm must swim through the water to reach the eggs is referred to as spawning.
Herbicide	Herbicide refers to a chemical substance that kills plants.
Biological control	Biological control of pests and diseases is a method of controlling pests and diseases in agriculture that relies on natural predation rather than introduced chemicals.
Pathogen	A pathogen or infectious agent is a biological agent that causes disease or illness to its host.The term is most often used for agents that disrupt the normal physiology of a multicellular animal or plant.
Geographic range	The geographic area in which a population lives is a geographic range.
Subsidence	Subsidence is a term used in geology, engineering and surveying to denote the motion of a surface (usually, the earth's surface) downwards relative to a datum such as sea-level.
Seedling	Seedling refers to a young plant that has grown from a seed. The most common example of germination is the sprouting of a seedling from a seed of a flowering plant or gymnosperm.
Hydrogen sulfide	The gas that is produced in anoxic sediments is referred to as hydrogen sulfide.
Carbon dioxide	Carbon dioxide is an atmospheric gas comprized of one carbon and two oxygen atoms. A very widely known chemical compound, it is frequently called by its formula CO_2. In its solid state, it is commonly known as dry ice.
Anaerobic	An anaerobic organism is any organism that does not require oxygen for growth.
Toxin	Toxin refers to a microbial product or component that can injure another cell or organism at low concentrations. Often the term refers to a poisonous protein, but toxins may be lipids and other substances.
Tannins	Tannins refers to polyphenolic compounds produced by most plants that bind proteins, thereby impairing digestion by herbivores and inhibiting microbes.
Shrub	A shrub or bush is a horticultural rather than strictly botanical category of woody plant, distinguished from a tree by its multiple stems and lower height, usually less than 6 m tall. A large number of plants can be either shrubs or trees, depending on the growing conditions they experience.
Pneumatophore	Pneumatophore refers to spongy root extending above the surface of the water, produced by a plant growing in water to facilitate oxygen absorption.
Fruit	A fruit is the ripened ovary—together with seeds—of a flowering plant. In many species, the fruit incorporates the ripened ovary and surrounding tissues.

Aerobic metabolism	Aerobic metabolism is rather more efficient than anaerobic metabolism. They share the initial pathway of glycolysis but aerobic metabolism continues with the Krebs cycle and oxidative phosphorylation. The post glycolytic reactions take place in the mitochondria in eukaryotic cells, and at the cell membrane in prokaryotic cells.
Crystal	Crystal is a solid in which the constituent atoms, molecules, or ions are packed in a regularly ordered, repeating pattern extending in all three spatial dimensions.
Circulatory system	The circulatory system or cardiovascular system is the organ system which circulates blood around the body of most animals.
Tropical forest	A tropical forest is a tropical and subtropical biome. They are moist broadleaf forests found in a belt around the equator and in the humid subtropics, and are characterized by warm, humid climates with high year-round rainfall.
Detritus feeders	Organisms, such as termites, fungi, and bacteria, that obtain their nutrients and energy mainly by feeding on dead organic matter are detritus feeders.
Isotope	An isotope is a form of an element whose nuclei have the same atomic number - the number of protons in the nucleus - but different mass numbers because they contain different numbers of neutrons.
Polymorphism	The presence in a population of more than one allele of a gene at a frequency greater than that of newly arising mutations is referred to as polymorphism.
Trophic structure	The different feeding relationships in an ecosystem, which determine the route of energy flow and the pattern of chemical cycling is called trophic structure.
Density dependence	Change in the severity of action of agents affecting birth and death rates within populations that are directly or inversely related to population density is density dependence.
Species diversity	Species diversity refers to a weighted representation of the species of organisms living in a region; large and common species are given greater weight than are small and rare ones.
Filter feeder	A Filter feeder is an animal that feeds by straining suspended matter and food particles from water, typically by passing the water over a specialized structure, such as the baleen of baleen whales.
Paleoecology	Paleoecology uses the data of fossils or subfossils to reconstruct the ecosystems of the past. It includes the study of fossil organisms in terms of their life cycle, their interactions, their natural environment, their manner of death and burial.
Fossil	A preserved remnant or impression of an organism that lived in the past is referred to as fossil.
Ecological succession	Ecological succession refers to the process of biological community change resulting from disturbance; transition in the species composition of a biological community, often following a flood, fire, or volcanic eruption.
Somatic cell	A somatic cell is generally taken to mean any cell forming the body of an organism.
Systematics	In biology, systematics is the study of the diversity of organism characteristics, and especially how they relate evolutionarily. The term "systematics" may or may not overlap with "taxonomy" which concerns itself with scientific classification of species and other taxa.
Perennial	Flowering plant that lives more than one growing season because the underground parts regrow each season is referred to as perennial.
Wetland	A wetland is an environment "at the interface between truly terrestrial ecosystems...and truly aquatic systems...making them different from each yet highly dependent on both."
Top-down control	Top-down control refers to an ecosystem in which the number of organisms in any trophic level

	is determined by the abundance of predators.
Ecosystem	In general terms an ecosystem can be thought of as an assemblage of organisms (plant, animal and other living organisms living together with their environment, functioning as a loose unit. That is, a dynamic and complex whole, interacting as an "ecological unit".
Growth form	One of several categories of the physical structure of plants, such as tree, herbaceous perennial, or liana is called growth form.
Biomass	Biomass is organic non-fossil material, collectively. In other words, biomass comprises the mass of all biological organisms, dead or alive, excluding biological mass that has been transformed by geological processes into substances such as coal or petroleum.
Botany	Botany is the scientific study of plant life. As a branch of biology, it is also sometimes referred to as plant science or plant biology.

Coral reef	A coral reef is a type of biotic reef that develops in tropical waters by the growth of coralline algae, hermatypic corals, and other marine organisms.
Kelp	Kelp are large seaweeds, belonging to the brown algae and classified in the order Laminariales. Despite their appearance they are not grouped with the normal aquatic or land plants (kingdom Plantae), but instead are included in either kingdom Protista or Chromista.
Primary producer	Primary producer refers to a photosynthetic or chemosynthetic organism that synthesizes complex organic molecules from simple inorganic ones.
Substrate	A substrate is a molecule which is acted upon by an enzyme. Each enzyme recognizes only the specific substrate of the reaction it catalyzes. A surface in or on which an organism lives.
Habitat	Habitat refers to a place where an organism lives; an environmental situation in which an organism lives.
Carbon	Carbon is a chemical element in the periodic table that has the symbol C and atomic number 6. An abundant nonmetallic, tetravalent element, carbon has several allotropic forms.
Microhabitat	A microhabitat is the immediate surroundings and other physical factors of an individual plant or animal within its habitat.
Rhizome	A rhizome is a usually-underground, horizontal stem of a plant that often sends out roots and shoots from its nodes.
Sediment	Sediment is any particulate matter that can be transported by fluid flow and which eventually is deposited as a layer of solid particles on the bed or bottom of a body of water or other liquid.
Biota	Biota is the plant and animal life of a region or area or geological period. The biota of the Earth is known as the biosphere.
Species	Group of similarly constructed organisms capable of interbreeding and producing fertile offspring is a species.
Temperate	In virology, temperate refers to the life cycle a phage is able to perform. The phage usually integrates its genome into its host chromosome, becoming a prophage.
Zostera	Zostera is a small genus of widely distributed seagrass, commonly called eelgrass. It contains twelve species. Zostera is found on sandy sunstrates or in esturaries submerged or partially floating. Most Zostera are perennial, they have long, bright green, ribbon-like leaves, about 1 cm wide.
Root	In vascular plants, the root is that organ of a plant body that typically lies below the surface of the soil. However, this is not always the case, since a root can also be aerial (that is, growing above the ground) or aerating (that is, growing up above the ground or especially above water).
Shoot	In botany, the shoot is one of two primary sections of a plant; the other is the root. The shoot refers to what is generally the upper portion of a plant, and consists of stems, leaves, flowers, and fruits. It is derived from the embryonic epicotyl, the portion of the embryo above the point of attachment to the seed leaves (cotyledons).
Flower	A flower is the reproductive structure of a flowering plant. The flower structure contains the plant's reproductive organs, and its function is to produce seeds through sexual reproduction.
Photosynthesis	Photosynthesis is a biochemical process in which plants, algae, and some bacteria harness the energy of light to produce food. Ultimately, nearly all living things depend on energy produced from photosynthesis for their nourishment, making it vital to life on Earth.

Meristem	Meristem is a type of embryonic tissue in plants consisting of unspecialized, youthful cells called meristematic cells and found in areas of the plant where growth is or will take place - the roots and shoots.
Epiphyte	The term epiphyte refers to any plant that grows upon or attached to another living plant.
Nitrogen-fixing bacteria	Microorganisms that restock nitrogenous minerals in the soil by converting nitrogen to ammonia are referred to as nitrogen-fixing bacteria.
Nitrogen	A colorless and tasteless and mostly inert diatomic non-metal gas that is an essential constituent of proteins is nitrogen.
Erosion	Erosion is the displacement of solids (soil, mud, rock, and other particles) by the agents of wind, water, ice, movement in response to gravity, or living organisms.
Colonization	Colonization is the act where life forms move into a distant area where their kind is sparse or not yet existing at all and set up new settlements in the area. Colonization applies to all life forms in a sense though it is most often used in reference to insects and humans.
Seaweed	Phycologists consider seaweed to refer any of a large number of marine benthic algae that are multicellular, macrothallic (large-bodied), and thus differentiated from most algae that tend to be microscopic in size.
Seedling	Seedling refers to a young plant that has grown from a seed. The most common example of germination is the sprouting of a seedling from a seed of a flowering plant or gymnosperm.
Salt marsh	Salt marsh refers to a grassy area that extends along the shores of estuaries and sheltered coasts in temperate and subpolar regions.
Bryozoan	Colonial animals that build stony skeletons of calcium carbonate and superficially resemble coral is referred to as bryozoan.
Sponge	An invertebrates that consist of a complex aggregation of cells, including collar cells, and has a skeleton of fibers and/or spicules is a sponge. They are primitive, sessile, mostly marine, waterdwelling filter feeders that pump water through their matrix to filter out particulates of food matter.
Invertebrate	Invertebrate is a term coined by Jean-Baptiste Lamarck to describe any animal without a spinal column. It therefore includes all animals except vertebrates (fish, reptiles, amphibians, birds and mammals).
Larva	A free-living, sexually immature form in some animal life cycles that may differ from the adult in morphology, nutrition, and habitat is called larva.
Byssal thread	Byssal thread denotes strong threads secreted by mussels to attach to rocks and large, generally heavy objects in the intertidal zone. They range to 6 centimeters in length.
Predator	A predator is an animal or other organism that hunts and kills other organisms for food in an act called predation.
Suspension feeder	An animal that extracts food particles suspended in the surrounding water for food is referred to as a suspension feeder.
Carnivore	An animal that eats a diet consisting solely of meat is referred to as a carnivore.
Snail	The name snail applies to most members of the molluscan class Gastropoda that have coiled shells. Other gastropods, which lack a conspicuous shell, are commonly called slugs, and are scattered throughout groups that primarily include snails.
Nutrient regeneration	Nutrient regeneration refers to the release of nutrients from organic matter by decomposers.

Decomposition	Decomposition refers to the reduction of the body of a formerly living organism into simpler forms of matter.
Diatom	Diatom refers to a unicellular photosynthetic alga with a unique, glassy cell wall containing silica.
Herbivore	A herbivore is an animal that is adapted to eat primarily plant matter
Sea urchin	A sea urchin is a spiny sea creature of the class Echinoidea found in oceans all over the world. Their shell, which biologists call the test, is globular in shape, and covered with spines.
Cellulose	A large polysaccharide composed of many glucose monomers linked into cable-like fibrils that provide structural support in plant cell walls is referred to as cellulose.
Diadema antillarum	Diadema antillarum is a sea urchin of the genus diadema with long black spines, it was once the most abundant and important herbivore on the coral reefs of the western Atlantic and Caribbean basin. In this role it was the main grazer that maintained the balance between algae growth and destruction.
Pollution	Any environmental change that adversely affects the lives and health of living things is referred to as pollution.
Catastrophe	An unpredictable event that has a strong effect on a population is referred to as a catastrophe.
Pathogen	A pathogen or infectious agent is a biological agent that causes disease or illness to its host.The term is most often used for agents that disrupt the normal physiology of a multicellular animal or plant.
Epidemic	An epidemic is a disease that appears as new cases in a given human population, during a given period, at a rate that substantially exceeds what is "expected", based on recent experience.
Salinity	Salinity refers to the total amount of salts dissolved in seawater. It is generally expressed in parts per thousand.
Phytoplankton	Phytoplankton refers to the autotrophic component of the plankton that drifts in the water column. Phytoplankton obtain energy through a process called photosynthesis, and so must live in the well-lit surface layer (termed the euphotic zone) of an ocean, sea, or lake.
Grazing	Grazing refers to consumption of a portion of a plant's tissues, generallyapplied to grasses and other herbaceous vegetation; removing plant or algal cover from a substrate, generally applied to aquatic systems.
Extinction	In biology and ecology, extinction is the ceasing of existence of a species or group of taxa. The moment of extinction is generally considered to be the death of the last individual of that species.The death of all members of a species is extinction.
Growth rate	A measure of the change in population size per individual per unit of time is called growth rate.
Gametophyte	Gametophyte refers to the multicellular haploid form in the life cycle of organisms undergoing alternation of generations; mitotically produces haploid gametes that unite and grow into the sporophyte generation.
Sporophyte	A sporophyte is the diploid structure or phase of life of a sexually reproducing plant. Each living cell of the sporophyte contains two complete sets of chromosomes. The sporophyte is the dominant life form in ferns, gymnosperms, and angiosperms.
Holdfast	A rootlike structure that anchors a seaweed is called holdfast.

Stipe	A stipe is a supportive structure that may be stemlike, as in seaweed, or a true leaf stem, as in ferns.
Stem	Stem refers to that part of a plant's shoot system that supports the leaves and reproductive structures.
Tissue	Group of similar cells which perform a common function is called tissue.
Canopy	Canopy refers to the leaf-bearing part of a tree. Collectively the aggregate of the leaves and branches of the larger woody plants of an ecological community.
Bulb	An underground food-storage organ that is essentially a modified bud consisting of fleshy leaves that surround and are attached to a small stem is called bulb.
Understory	The aggregate of smaller plants growing beneath the canopy of dominant plants in a forest is an understory.
Mussel	A mussel is a bivalve mollusk that can be found in lakes, rivers, creeks, intertidal areas, and throughout the ocean. The saltwater mussels (family Mytilidae) and freshwater mussels (family Unionidae) are not thought to be closely related and are grouped in different subclasses, despite considerable similarities in appearance.
Mollusk	The mollusk is the large and diverse phylum Mollusca, which includes a variety of familiar creatures well-known for their decorative shells or as seafood. These range from tiny snails, clams, and abalone to the octopus, cuttlefish and squid (which are considered the most intelligent invertebrates).
Limpet	Limpet is a marine mollusk in the family Acmaeidae with flattened, cone-shaped shells. They live throughout the intertidal zone, from the high zone (upper littoral) to the shallow subtidal on the rocky coasts of most oceans. They can be commonly found attached to rocks, looking like little disks or bumps on the rock surface.
Tube feet	Tube feet are the many small tubular projections found most famously on the ventral face of a sea star's arms, but are characteristic of the water vascular system of the echinoderm phylum which also includes sea urchins, sand dollars and sea cucumbers and many other sea creatures.
Dorsal	In anatomy, the dorsal is the side in which the backbone is located. This is usually the top of an animal, although in humans it refers to the back.
Spine	A spine is a rigid, pointed surface protuberance or needle-like structure on an animal, shell, or plant, presumably serving as a defense against attack by predators.
Population	Group of organisms of the same species occupying a certain area and sharing a common gene pool is referred to as population.
Browser	A type of herbivore whose nutrition generally comes from high growing plants, like trees, rather than a grazer that eats from the ground is a browser. Deer and goats are domesticated browsers, and elephants and giraffes are wild browsers.
Zoospore	Zoospore refers to a nonsexual reproductive cell that swims by using flagella; formed by members of the protistan division Oomycota.
Haploid	Haploid cells bear one copy of each chromosome.
Fusion	Fusion refers to the combination of two atoms into a single atom as a result of a collision, usually accompanied by the release of energy.
Egg	An egg is the zygote, resulting from fertilization of the ovum. It nourishes and protects the embryo.
Genus	In biology, a genus is a taxonomic grouping. That is, in the classification of living organisms, a genus is considered to be distinct from other such genera. A genus has one or

	more species: if it has more than one species these are likely to be morphologically more similar than species belonging to different genera.
Sunlight	Sunlight in the broad sense is the total spectrum of electromagnetic radiation given off by the Sun.
Algae	The algae consist of several different groups of living organisms that capture light energy through photosynthesis, converting inorganic substances into simple sugars with the captured energy.
Barren	An area with sparse vegetation owing to some physical or chemical property of the soil is referred to as barren.
Dorsal surface	The dorsal surface is arched from above downward, and is subdivided into two unequal parts by the spine; the portion above the spine is called the supraspinatous fossa, and that below it the infraspinous fossa.
Threshold	Electrical potential level at which an action potential or nerve impulse is produced is called threshold.
Trophic structure	The different feeding relationships in an ecosystem, which determine the route of energy flow and the pattern of chemical cycling is called trophic structure.
Population density	The number of organisms of a species per unit area is called population density.
Trophic level	In ecology, the trophic level is the position that an organism occupies in a food chain - what it eats, and what eats it.
Starfish	Starfish are marine invertebrates belonging to phylum Echinodermata, class Asteroidea. They exhibit a superficially radial symmetry, typically with five or more "arms" which radiate from an indistinct disk. They do not have movable skeletons, but instead possess a hydraulic water vascular system.
Perennial	Flowering plant that lives more than one growing season because the underground parts regrow each season is referred to as perennial.
Calcareous	Calcareous refers to a sediment, sedimentary rock, or soil type which is formed from or contains a high proportion of calcium carbonate in the form of calcite or aragonite. It is also used to refer to relatively alkaline soil.
Coral	Group of cnidarians having a calcium carbonate skeleton that participate in the formation of reefs is coral.
Calcium carbonate	A mineral that is the major component of the shell, skeleton, and other parts of many organisms are called calcium carbonate.
Calcareous algae	Calcareous algae is algae that deposits calcium carbonate (limestone) in its tissue. When the algae dies, it leaves a fossil skeleton behind. Most calcareous algae are red, but there are some exceptions.
Skeleton	In biology, the skeleton or skeletal system is the biological system providing physical support in living organisms.
Barnacle	A barnacle is a type of arthropod belonging to infraclass Cirripedia in the subphylum Crustacea and is hence distantly related to crabs and lobsters. They live attached to surfaces and are typically enclosed by heavy calcareous plates are referred to as barnacles.
Sea squirts	Tunicates with a sac-like, attached body called a tunic as adults are referred to as sea squirts.
Bivalve	Bivalve refers to a member of a group of mollusks that includes clams, mussels, scallops, and

	oysters. They typically have two-part shells, with both parts being more or less symmetrical.
Dominance	Dominance is the state of having high social status relative to other individuals, who react submissively to dominant individuals.
Intertidal zone	The part of a beach that lies between average high tide and average low tide is called the intertidal zone.
Sessile	In botany, sessile means without a stalk, as in flowers or leaves that grow directly from the stem. In zoology, sessile organisms are those which are not able to move about. They are usually permanently attached to a solid substrate of some kind, such as a rock, or the hull of a ship in the case of barnacles.
Surface layer	The upper layer of water that is mixed by wind, waves, and currents is the surface layer .
Surface water	Includes all bodies of water, lakes, rivers, ponds, and so on that are on the surface of the earth, in contrast to groundwater, which lies below the surface is surface water.
Modular organism	A modular organism grows by producing additional units of body construction that are very similar to the units of which it is already composed.
Polyp	Small, abnormal growth that arises from the epithelial lining is referred to as polyp. In zoology, a polyp is one of two forms of individuals found in many species of cnidarians. They are approximately cylindrical, elongated on the axis of the body.
Zooid	One of the distinct individuals forming a colonial animal such as a bryozoan or hydrozoan is called zooid.
Predation	Interaction in which one organism uses another, called the prey, as a food source is referred to as predation.
Colony	Colony refers to a cluster or assemblage of microorganisms growing on the surface of an agar culture medium. A colony also refers to several individual organisms of the same species living closely together, usually for mutual benefit, such as stronger defences, the ability to attack bigger prey, etc.
Red algae	One of a group of marine, mostly multicellular, autotrophic protists, which includes the reef-building coralline algae is referred to as red algae.
Intraspecific competition	Competition between individuals of the same species in a community for a limited resource is referred to as intraspecific competition.
Distribution	Distribution in pharmacology is a branch of pharmacokinetics describing reversible transfer of drug from one location to another within the body.
Dispersal	Biological dispersal refers to those processes by which a species maintains or expands the distribution of a population. Dispersal implies movement—movement away from an existing population or away from the parent organisms.
Biology	Biology is the branch of science dealing with the study of life. It is concerned with the characteristics, classification, and behaviors of organisms, how species come into existence, and the interactions they have with each other and with the environment.
Fauna	Fauna (animals), a collective term for animal life, as distinct from flora (plant life).
Biotic	Biotic refers to life or living organisms.
Closed system	An ecosystem that can can exchange heat and work, but not matter, with its surroundings is called a closed system.
Tadpole	A tadpole is a larval frog, toad, salamander, newt, or caecilian. In this stage it breathes by means of external or internal gills, is at first lacking legs, and has a finlike tail with

	which it swims as most fish do, by lateral undulation.
Metapopulation	A metapopulation consists of a group of spatially separated populations of the same species which interact at some level.
Dominant species	Dominant species refers to those species in a community that have the highest abundance or highest biomass. These species exert a powerful control over the occurrence and distribution of other species.
Morphology	The scientific study of organic form, including both its development and function is morphology.
Secretion	Secretion is the process of segregating, elaborating, and releasing chemicals from a cell, or a secreted chemical substance or amount of substance.
Coexistence	Coexistence refers to occurrence of two or more species in the same habitat; usually applied to potentially competing species.
Downstream	Downstream refers to a relative position in DNA or RNA. Relative to the position on the strand, downstream is the region towards the 3' end of the strand.
Upstream	Upstream refers to movement opposite the direction RNA follows when moving along a gene.
Ecosystem	In general terms an ecosystem can be thought of as an assemblage of organisms (plant, animal and other living organisms living together with their environment, functioning as a loose unit. That is, a dynamic and complex whole, interacting as an "ecological unit".
Linkage	Linkage refers to the patterns of assortment of genes that are located on the same chromosome. Important because if the genes are located relatively far apart, crossing over is more likely to occur between them than if they are close together.
Food chain	Food chain refers to a sequence of food transfers from one species to another within an ecosystem.
Biomass	Biomass is organic non-fossil material, collectively. In other words, biomass comprises the mass of all biological organisms, dead or alive, excluding biological mass that has been transformed by geological processes into substances such as coal or petroleum.
Lagoon	A lagoon is a body of comparatively shallow salt water separated from the deeper sea by a shallow or exposed sandbank, coral reef, or similar feature. Thus, the enclosed body of water behind a barrier reef or barrier islands or enclosed by an atoll reef is called a lagoon.
Pinniped	Pinniped refers to mammals with paddle-shaped flippers: seals, eared seals, and the walrus.
Food web	A network of interconnecting food chains is referred to as a food web.
Fjord	A fjord is a narrow inlet of the sea between cliffs or steep slopes, which results from marine inundation of a glaciated valley. Typical characteristics of a fjord include: a narrow inlet, a bottom glacially eroded significantly below sea level, steep-sided walls which continue to descend below the sea surface, greater depths in the upper and middle reaches than on the seaward side, and communication with the open sea.
Zooxanthellae	Zooxanthellae refers to dinoflagellates and other algae that live as symbionts within coral.
Matrix	In biology, matrix (plural: matrices) is the material between animal or plant cells, the material (or tissue) in which more specialized structures are embedded, and a specific part of the mitochondrion that is the site of oxidation of organic molecules.
Coralline algae	Red algae that deposit calcium carbonate in their thallus are called the coralline algae.
Silica	Silica is a mineral similar to glass that is the major component of the cell wall, shell, or skeleton of many marine organisms.

Cnidaria	Cnidaria is a phylum containing some 10,000 species of relatively simple animals found exclusively in aquatic environments (most species are marine). They get their name from cnidocytes, which are specialized cells that carry stinging organelles.
Phylum	Phylum is the second highest taxonomic classification for the kingdom Animalia (animals), between kingdom level and class level.
Tentacle	An elongate, extensible projection of the body of cnidarians and cephalopod mollusks that may be used for grasping, stinging, and immobilizing prey, and for locomotion is called tentacle.
Zooplankton	Nonphotosynthetic protists that feed on other plankton and are abundant in marine and freshwater environments are zooplankton.
Nematocyst	In cnidarians, a capsule that contains a threadlike fiber whose release aids in the capture of prey is referred to as nematocyst.
Nutrition	Nutrition refers to collectively, the processes involved in taking in, assimilating, and utilizing nutrients.
Symbiosis	Symbiosis is an interaction between two organisms living together in more or less intimate association or even the merging of two dissimilar organisms. The term host is usually used for the larger (macro) of the two members of a symbiosis.
Mutualism	Mutualism is an interaction between two species in which both species derive benefit.
Dinoflagellate	A unicellular photosynthetic alga with two flagella situated in perpendicular grooves in cellulose plates covering the cell is a dinoflagellate.
Sea anemone	Named after a terrestrial flower, the sea anemone is a group of water-dwelling, filter feeding animals of the order Actinaria. As a cnidarian, it is closely related to coral and jellyfish.
Excretion	Excretion is the biological process by which an organism chemically separates waste products from its body. The waste products are then usually expelled from the body by elimination.
Carbohydrate	Carbohydrate is a chemical compound that contains oxygen, hydrogen, and carbon atoms. They consist of monosaccharide sugars of varying chain lengths and that have the general chemical formula $C_n(H_2O)_n$ or are derivatives of such.
Radiation	The emission of electromagnetic waves by all objects warmer than absolute zero is referred to as radiation.
Amino acid	An amino acid is any molecule that contains both amino and carboxylic acid functional groups. They are the basic structural building units of proteins. They form short polymer chains called peptides or polypeptides which in turn form structures called proteins.
Protein	A protein is a complex, high-molecular-weight organic compound that consists of amino acids joined by peptide bonds. They are essential to the structure and function of all living cells and viruses. Many are enzymes or subunits of enzymes.
Soft coral	Soft coral is a group of coral species that are thick and fleshy, and that move and wave in ocean water currents. These corals do not produce calcium carbonate skeleton; they include leather coral, cauliflower coral, and tree coral.
Calcification	A process in which tissue becomes hardened due to deposits of calcium salts. Calcification of blood vessels plays a role in the development of atherosclerosis.
Radioactive	A term used to describe the property of releasing energy or particles from an unstable atom is called radioactive.
Calcium	Calcium is the chemical element in the periodic table that has the symbol Ca and atomic number 20. Calcium is a soft grey alkaline earth metal that is used as a reducing agent in

	the extraction of thorium, zirconium and uranium. Calcium is also the fifth most abundant element in the Earth's crust.
Isotope	An isotope is a form of an element whose nuclei have the same atomic number - the number of protons in the nucleus - but different mass numbers because they contain different numbers of neutrons.
Inhibitor	An inhibitor is a type of effector (biology) that decreases or prevents the rate of a chemical reaction. They are often called negative catalysts.
Carbon dioxide	Carbon dioxide is an atmospheric gas comprized of one carbon and two oxygen atoms. A very widely known chemical compound, it is frequently called by its formula CO_2. In its solid state, it is commonly known as dry ice.
Lipid	Lipid is one class of aliphatic hydrocarbon-containing organic compounds essential for the structure and function of living cells. They are characterized by being water-insoluble but soluble in nonpolar organic solvents.
Acetate	Acetate is the anion of a salt or ester of acetic acid.
Ultraviolet radiation	Ultraviolet radiation refers to radiation similar to light but with wavelengths slightly shorter than violet light and with more energy. The greater energy causes it to severely burn and otherwise damage biological tissues.
Pigment	Pigment is any material resulting in color in plant or animal cells which is the result of selective absorption.
Bacteria	The domain that contains procaryotic cells with primarily diacyl glycerol diesters in their membranes and with bacterial rRNA. Bacteria also is a general term for organisms that are composed of procaryotic cells and are not multicellular.
Brain	The part of the central nervous system involved in regulating and controlling body activity and interpreting information from the senses transmitted through the nervous system is referred to as the brain.
Mucus	Mucus is a slippery secretion of the lining of various membranes in the body (mucous membranes). Mucus aids in the protection of the lungs by trapping foreign particles that enter the nose during normal breathing. Additionally, it prevents tissues from drying out.
Reef crest	Reef crest refers to the shallow outer edge of the reef slope of a coral reef .
Isthmus	An isthmus is a narrow strip of land that is bordered on two sides by water and connects two larger land masses. It is the inverse of a strait (which lies between two land masses and connects two larger bodies of water).
Pleistocene	A geologic epoch, characterized by alternating glacial and interglacial stages, that ended about 10,000 years ago, that lasted for 2 million years is called the pleistocene epoch.
Atoll	An atoll is a type of low, coral island found in tropical oceans and consisting of a coral-algal reef surrounding a central depression. The depression may be part of the emergent island, but more typically is a part of the sea (that is, a lagoon), or very rarely is an enclosed body of fresh, brackish, or highly saline water.
Cone	Cone refers to a reproductive structure of gymnosperms that produces pollen in males or eggs in females.
Charles Darwin	Charles Darwin was a British naturalist who achieved lasting fame as originator of the theory of evolution through natural selection.
Barrier reef	A type of coral reef that develops at some distance from the coast creating a lagoon between itself and the coast is called a barrier reef.

Zonation	The distribution of organisms in bands or regions corresponding to changes in ecological conditions along a continuum, for example, intertidal zonation and elevational zonation.
Sea cucumber	The sea cucumber is an echinoderm of the class Holothuroidea, with an elongated body and leathery skin. Mostly found on the sea floor.
Water column	The vertical column of seawater that extends from the surface to the bottom is called water column.
Spawning	A method of external fertilization in which male and female parents shed gametes into the water, and sperm must swim through the water to reach the eggs is referred to as spawning.
Gamete	A gamete is a specialized germ cell that unites with another gamete during fertilization in organisms that reproduce sexually. They are haploid cells; that is, they contain one complete set of chromosomes. When they unite they form a zygote—a cell having two complete sets of chromosomes and therefore diploid.
Simultaneous hermaphrodite	Simultaneous hermaphrodite is when the organism has both male and female sexual organs at the same time as an adult. Usually, self-fertilization does not occur.
Sperm	Sperm refers to the male sex cell with three distinct parts at maturity: head, middle piece, and tail.
Fertilization	Fertilization is fusion of gametes to form a new organism. In animals, the process involves a sperm fusing with an ovum, which eventually leads to the development of an embryo.
Neap tides	Neap tides refers to the tides with a small tidal range. Neap tides occur when the gravitational forces of the Moon and the Sun are perpendicular to one another and occur during quarter moons.
Adaptation	A biological adaptation is an anatomical structure, physiological process or behavioral trait of an organism that has evolved over a period of time by the process of natural selection such that it increases the expected long-term reproductive success of the organism.
Gene	Gene refers to a discrete unit of hereditary information consisting of a specific nucleotide sequence in DNA . Most of the genes of a eukaryote are located in its chromosomal DNA; a few are carried by the DNA of mitochondria and chloroplasts.
Genetic drift	Genetic drift is the term used in population genetics to refer to the statistical drift over time of allele frequencies in a population due to random sampling effects in the formation of successive generations.
Natural selection	Natural selection is the process by which biological individuals that are endowed with favorable or deleterious traits end up reproducing more or less than other individuals that do not possess such traits.
Allele	An allele is any one of a number of viable DNA codings of the same gene (sometimes the term refers to a non-gene sequence) occupying a given locus (position) on a chromosome.
Biological interaction	A biological interaction results from the fact that organisms in an ecosystem interact with each other, in the natural world, no organism is an autonomous entity isolated from its surroundings.
Digestion	Digestion refers to the mechanical and chemical breakdown of food into molecules small enough for the body to absorb; the second main stage of food processing, following ingestion.
Filament	The stamen is the male organ of a flower. Each stamen generally has a stalk called the filament, and, on top of the filament, an anther. The filament is a long chain of proteins, such as those found in hair, muscle, or in flagella.
Sweeper tentacle	A sweeper tentacle is an elongated tentacle equipped with stinging nematocyst cells found

	mainly on large polyped stony corals and used for defence.
Cell	The cell is the structural and functional unit of all living organisms, and is sometimes called the "building block of life."
Genotype	The genotype is the specific genetic makeup (the specific genome) of an individual, usually in the form of DNA. It codes for the phenotype of that individual.
Chimera	An embryo or animal composed of cells from two or more different organisms is a chimera.
Keystone species	Species that are not usually abundant in a community yet exert strong control on community structure by the nature of their ecological roles or niches are keystone species.
Eon	An eon is composed of several eras, which in turn are composed of periods, which are composed of epochs. One definition of an eon puts it as "a unit of geologic time equal to one billion years", although it is rarely used in this manner.
Symbiotic relationship	Symbiotic relationship refers to an interspecific interaction in which one species, the symbiont, lives in or on another species, the host; a close association between organisms of two or more species.
Gastropod	A member of the largest group of mollusks, including snails and slugs is a gastropod.
Lizard	A lizard is a reptile of the order Squamata, which they share with the snakes (Ophidians). They are usually four-legged, with external ear openings and movable eyelids. Species range in adult length from a few centimeters (some Caribbean geckos) to nearly three meters (Komodo dragons).
Sclerosponge	Sclerosponge is a sponge with a soft body that covers a hard, often massive skeleton made of calcium carbonate, either aragonite or calcite.
Deep sea	The deep sea refers to those areas of oceans to which little or no light penetrates.
Sulfur	Sulfur is the chemical element in the periodic table that has the symbol S and atomic number 16. It is an abundant, tasteless, odorless, multivalent non-metal. Sulfur, in its native form, is a yellow crystaline solid. In nature, it can be found as the pure element or as sulfide and sulfate minerals.
Host	Host is an organism that harbors a parasite, mutual partner, or commensal partner; or a cell infected by a virus.
Parasite	A parasite is an organism that spends a significant portion of its life in or on the living tissue of a host organism and which causes harm to the host without immediately killing it. They also commonly show highly specialized adaptations allowing them to exploit host resources.
Relative abundance	Differences in the abundance of species within a community is relative abundance.
Bleaching	Bleaching is a phenomenon where severely stressed corals turn white, frequently due to high water temperatures. This color change is mainly due to the corals expelling their zooxanthellae.
Oxygen	Oxygen is a chemical element in the periodic table. It has the symbol O and atomic number 8. Oxygen is the second most common element on Earth, composing around 46% of the mass of Earth's crust and 28% of the mass of Earth as a whole, and is the third most common element in the universe.
Overfishing	Overfishing is a situation where one or more fish stocks are reduced below predefined levels of acceptance by fishing activities. Biological overfishing occurs when fishing mortality has reached a level where the stock biomass has negative marginal growth .

Speciation	Speciation refers to the evolutionary process by which new biological species arise. All forms of speciation have actually taken place over the course of evolution, though it still remains a subject of debate as to the relative importance of each mechanism in driving biodiversity.
Gene flow	The gain or loss of alleles from a population by the movement of individuals or gametes into or out of the population is called gene flow.
Phenotypic plasticity	The ability of a genotype to change its phenotype in response to changes in the environment is called phenotypic plasticity.
Growth form	One of several categories of the physical structure of plants, such as tree, herbaceous perennial, or liana is called growth form.
Biodiversity	Biodiversity is the totality of genes, species, and ecosystems of a region.
Polymorphism	The presence in a population of more than one allele of a gene at a frequency greater than that of newly arising mutations is referred to as polymorphism.
Enzyme	An enzyme is a protein that catalyzes, or speeds up, a chemical reaction. They are essential to sustain life because most chemical reactions in biological cells would occur too slowly, or would lead to different products, without them.
Reproductive isolation	Reproductive isolation refers to the failure of organisms of one population to breed successfully with members of another; may be due to premating or postmating isolating mechanisms.
DNA markers	DNA markers refers to a purified fragment of DNA labeled with a radioactive isotope or fluorescent dye and used to identify complementary sequences by means of hybridization.
Isolating mechanism	A morphological, physiological, behavioral, or ecological difference that prevents members of two species from interbreeding is referred to as an isolating mechanism.
Hybridization	In molecular biology hybridization is the process of joining two complementary strands of
Commensal	Living on or within another organism without injuring or benefiting the other organism is referred to as commensal.
Crab	The term crab is often applied to several different groups of short (nose to tail) decapod crustaceans with thick exoskeletons, but only members of the Brachyura are true.
Excavates	The excavates are a major line of protists, including a variety of free-living and symbiotic forms. They are usually characterized by having two, four, or more flagella with distinct ultrastructure anterior to a ventral feeding groove supported by microtubules.
Antennae	Antennae are paired appendages connected to the front-most segments of arthropods. In crustaceans, they are biramous and present on the first two segments of the head, with the smaller pair known as antennules. In insects, olfactory receptors on the antennae bind to odor molecules, including pheromones.
Territorial behavior	Behavior involved in establishing, defending, and maintaining a territory for food, mating, or other purposes is called territorial behavior.
Ectoparasite	A parasite that lives on or attached to the host's surface is referred to as ectoparasite.
Sulfuric acid	Sulfuric acid refers to the major constituent of acid precipitation. Formed as a result of sulfur dioxide emissions reacting with water vapor in the atmosphere.
Tunicate	One of a group of invertebrate chordates in the subphylum of saclike filter feeders with input and output siphons is referred to as a tunicate.
Vacuole	A vacuole is a large membrane-bound compartment within some eukaryotic cells where they serve

a variety of different functions: capturing food materials or unwanted structural debris surrounding the cell, sequestering materials that might be toxic to the cell, maintaining fluid balance (called turgor) within the cell.

Acid — An acid is a water-soluble, sour-tasting chemical compound that when dissolved in water, gives a solution with a pH of less than 7.

Prostaglandin — A prostaglandin is any member of a group of lipid compounds that are derived from fatty acids and have important functions in the animal body.

Hormone — A hormone is a chemical messenger from one cell to another. All multicellular organisms produce hormones. The best known hormones are those produced by endocrine glands of vertebrate animals, but hormones are produced by nearly every organ system and tissue type in a human or animal body. Hormone molecules are secreted directly into the bloodstream, they move by circulation or diffusion to their target cells, which may be nearby cells in the same tissue or cells of a distant organ of the body.

Gorgonians — Colonial anthozoans that secrete a skeleton made of protein are referred to as gorgonians. They are horny corals, an example is a Sea Fan.

Seagrass — Seagrass are flowering plants from four plant families (Posidoniaceae, Zosteraceae, Hydrocharitaceae and Cymodoceaceae) that grow in the marine saline environment.

Interstitial water — The water contained between sediment particles is interstitial water.

Rhizosphere — Rhizosphere is the zone of soil that is directly influenced by roots and associated soil microorganisms. This effect is by transfer of root exudates and root tissue to soil. This process is termed rhizodeposition.

Nitrate — Nitrate refers to a salt of nitric acid; a compound containing the radical NO_3; biologically, the final form of nitrogen from the oxidation of organic nitrogen compounds.

Ecology — Ecology is the scientific study of the distribution and abundance of living organisms and how these properties are affected by interactions between the organisms and their environment.

Phosphorus — Phosphorus is the chemical element in the periodic table that has the symbol P and atomic number 15.

Benthos — Benthos are the organisms and habitats of the sea floor; in freshwater biology they are the organisms and habitats of the bottoms of lakes, rivers, and creeks.

Population structure — Population structure refers to the proportions of individuals in a population belonging to different age classes. Also, the distribution of the population in space.

Systematics — In biology, systematics is the study of the diversity of organism characteristics, and especially how they relate evolutionarily. The term "systematics" may or may not overlap with "taxonomy" which concerns itself with scientific classification of species and other taxa.

Mosaic — An organism containing tissues of different genotypes is referred to as mosaic.

Secondary production — Secondary production is the biomass production resulting from the assimilation of organic matter produced by a primary consumer; production by organisms (mainly animals) which consume primary producers (mainly plants).

Detritus — Detritus is organic waste material from decomposing dead plants or animals.

Spore — Spore refers to a differentiated, specialized form that can be used for dissemination, for survival of adverse conditions because of its heat and dessication resistance, and/or for reproduction. They are usually unicellular and may develop into vegetative organisms or gametes. A spore may be produced asexually or sexually and are of many types.

Subtidal zone	The subtidal zone is the area below the range of tidal exposure; always covered by water.
Pinna	The pinna is the visible part of the ear that resides outside of the head. It acts as a funnel, amplifying the sound and directing it to the ear canal. While reflecting from the pinna, sound also goes through a filtering process which adds directional information to the sound.
Species diversity	Species diversity refers to a weighted representation of the species of organisms living in a region; large and common species are given greater weight than are small and rare ones.
Diffusion	Diffusion refers to the spontaneous movement of particles of any kind from where they are more concentrated to where they are less concentrated.
Echinodermata	Echinodermata refers to the phylum of echinoderms, including sea stars, sea urchins, and sand dollars; characterized by a rough or spiny skin, a water vascular system, an endoskeleton, and radial symmetry in adults.
Runoff	Water entering rivers, lakes, reservoirs, or the ocean from land surfaces is runoff.
Echinoderm	Echinoderm is a phylum of marine animals found in the ocean at all depths. This phylum dates back to the lower Cambrian period and represents about 7000 living species and 13000 extinct ones. 6 classes made it to the Modern Era.
Anthozoa	Anthozoa is a class within the phylum Cnidaria that contains the Sea anemones and corals. Unlike other cnidarians, anthozoans do not have a medusa stage in their development. Instead, they release sperm and eggs that form a planula, which attaches to some substrate on which the cnidarian grows.
Bioenergetics	The study of how organisms manage their energy resources is called bioenergetics.
Species richness	Species richness is perhaps the simplest measure of biodiversity. The larger the number, the more species in an area. It is referred to in equations as S. There is a strong inverse correlation in many groups between species richness and latitude - the further from the equator you go, the fewer species you will find.
Insect	An arthropod that usually has three body segments , three pairs of legs, and one or two pairs of wings is called an insect. They are the largest and (on land) most widely-distributed taxon within the phylum Arthropoda. They comprise the most diverse group of animals on the earth, with around 925,000 species described
Physiology	The study of the function of cells, tissues, and organs is referred to as physiology.
Paleontology	Paleontology refers to the scientific study of fossils and all aspects of extinct life.
Reproduction	Biological reproduction is the biological process by which new individual organisms are produced. Reproduction is a fundamental feature of all known life; each individual organism exists as the result of reproduction by an antecedent.
Respiration	Respiration is the process by which an organism obtains energy by reacting oxygen with glucose to give water, carbon dioxide and ATP (energy). Respiration takes place on a cellular level in the mitochondria of the cells and provide the cells with energy.
Microsatellite	A microsatellite is a short block of DNA sequence (a Tandemly Repetitive DNA sequence), often less than 150 base pairs long, that is repeated many times within the genome of an organism. Many repeats tend to be concentrated at the same locus.
Primary productivity	Primary productivity is the amount of light energy converted to chemical energy in a given amount of time. It usually corresponds to the amount of photosynthesis that occurred.
Echinoidea	Sea urchins are spiny sea creatures of the class Echinoidea found in oceans all over the world. (The name sea urchin means sea hedgehog, hedgehog being one meaning of the word

urchin). Their shell, which biologists call the test, is globular in shape, and covered with spines. The size of the test in adults is typically in the range of 3 to 10 cm.

Mating system The pattern of matings between individuals in a population, including number of simultaneous mates, permanence of pair bond, and degree of inbreeding is referred to as the mating system.

Continental shelf	The shallow part of the seafloor immediately adjacent to a continent is called the continental shelf.
Deep sea	The deep sea refers to those areas of oceans to which little or no light penetrates.
Habitat	Habitat refers to a place where an organism lives; an environmental situation in which an organism lives.
Sediment	Sediment is any particulate matter that can be transported by fluid flow and which eventually is deposited as a layer of solid particles on the bed or bottom of a body of water or other liquid.
Gradient	Gradient refers to a difference in concentration, pressure, or electrical charge between two regions.
Surface water	Includes all bodies of water, lakes, rivers, ponds, and so on that are on the surface of the earth, in contrast to groundwater, which lies below the surface is surface water.
Suspension feeder	An animal that extracts food particles suspended in the surrounding water for food is referred to as a suspension feeder.
Deposit feeder	A deposit feeder is an organism that feeds on the particles of matter in the soil, usually the top sediment where it is filled with organic matter.
Distribution	Distribution in pharmacology is a branch of pharmacokinetics describing reversible transfer of drug from one location to another within the body.
Carnivore	An animal that eats a diet consisting solely of meat is referred to as a carnivore.
Diatom	Diatom refers to a unicellular photosynthetic alga with a unique, glassy cell wall containing silica.
Clay	A soil constituent comprising particles smaller than 2 micrometers in diameter is clay. Clay is a generic term for an aggregate of hydrous silicate particles less than 2 ìm (micrometres) in diameter.
Phytoplankton	Phytoplankton refers to the autotrophic component of the plankton that drifts in the water column. Phytoplankton obtain energy through a process called photosynthesis, and so must live in the well-lit surface layer (termed the euphotic zone) of an ocean, sea, or lake.
Fecal	Fecal is the solid excretory wastes of animals. Consists of undigested material passing through the gut and bacteria that have begun to feed on it.
Invertebrate	Invertebrate is a term coined by Jean-Baptiste Lamarck to describe any animal without a spinal column. It therefore includes all animals except vertebrates (fish, reptiles, amphibians, birds and mammals).
Amensalism	Amensalism is an biological interaction, a type of symbiosis, between two species in which one impedes or restricts the success of the other without being affected, positively or negatively, by the presence of the other.
Organ	Organ refers to a structure consisting of several tissues adapted as a group to perform specific functions.
Water column	The vertical column of seawater that extends from the surface to the bottom is called water column.
Tissue	Group of similar cells which perform a common function is called tissue.
Coral reef	A coral reef is a type of biotic reef that develops in tropical waters by the growth of coralline algae, hermatypic corals, and other marine organisms.
Mangrove	Mangrove are woody trees or shrubs that grow in coastal habitats or mangal (Hogarth, 1999),

	for which the term mangrove swamp also would apply.
Mudflat	Mudflat is a coastal wetland that forms when mud is deposited by the tides or rivers, they are found in sheltered areas such as bays, bayous, lagoons, and estuaries.
Homogeneous	Homogeneous refers both to animals and plants, of having a resemblance in structure, due to descent from a common progenitor with subsequent modification.
Predator	A predator is an animal or other organism that hunts and kills other organisms for food in an act called predation.
Sea cucumber	The sea cucumber is an echinoderm of the class Holothuroidea, with an elongated body and leathery skin. Mostly found on the sea floor.
Brittle star	A brittle star is an echinoderm with five flexible arms that radiate from a conspicuous central disk and tube feet that are used in feeding.
Cone	Cone refers to a reproductive structure of gymnosperms that produces pollen in males or eggs in females.
Population	Group of organisms of the same species occupying a certain area and sharing a common gene pool is referred to as population.
Species	Group of similarly constructed organisms capable of interbreeding and producing fertile offspring is a species.
Amphipods	Amphipods include over 7000 described species of small, shrimp-like crustaceans. Most amphipods are marine, although a few live in freshwater or are terrestrial. One species common to mid-Atlantic and estuarine waters is Jassa falcata.
Detritus feeders	Organisms, such as termites, fungi, and bacteria, that obtain their nutrients and energy mainly by feeding on dead organic matter are detritus feeders.
Detritus	Detritus is organic waste material from decomposing dead plants or animals.
Continental rise	The gently sloping area at the base of the continental slope is called the continental rise.
Boundary layer	Boundary layer refers to a layer of still or slow-moving water or air close to the surface of an object.
Erosion	Erosion is the displacement of solids (soil, mud, rock, and other particles) by the agents of wind, water, ice, movement in response to gravity, or living organisms.
Annelid	Annelid refers to member of a phylum of invertebrates that contains segmented worms, such as the earthworm and the clam worm.
Copepods	A group of small crustaceans that are important members of marine and freshwater plankton communities is copepods.
Copepod	The group of small crustaceans that is an important member of marine and freshwater plankton communities is copepod. They are very important food organisms for small fish, whales, and other crustaceans such as krill in the ocean and in fresh water.
Decomposition	Decomposition refers to the reduction of the body of a formerly living organism into simpler forms of matter.
Oxygen	Oxygen is a chemical element in the periodic table. It has the symbol O and atomic number 8. Oxygen is the second most common element on Earth, composing around 46% of the mass of Earth's crust and 28% of the mass of Earth as a whole, and is the third most common element in the universe.
Nutrition	Nutrition refers to collectively, the processes involved in taking in, assimilating, and utilizing nutrients.

Bacteria	The domain that contains procaryotic cells with primarily diacyl glycerol diesters in their membranes and with bacterial rRNA. Bacteria also is a general term for organisms that are composed of procaryotic cells and are not multicellular.
Dispersal	Biological dispersal refers to those processes by which a species maintains or expands the distribution of a population. Dispersal implies movement—movement away from an existing population or away from the parent organisms.
Substratum	Substratum is the underlying layer, or base to which a lichen is fixed.
Predation	Interaction in which one organism uses another, called the prey, as a food source is referred to as predation.
Salinity	Salinity refers to the total amount of salts dissolved in seawater. It is generally expressed in parts per thousand.
Sea urchin	A sea urchin is a spiny sea creature of the class Echinoidea found in oceans all over the world. Their shell, which biologists call the test, is globular in shape, and covered with spines.
Mosaic	An organism containing tissues of different genotypes is referred to as mosaic.
Colonization	Colonization is the act where life forms move into a distant area where their kind is sparse or not yet existing at all and set up new settlements in the area. Colonization applies to all life forms in a sense though it is most often used in reference to insects and humans.
Dominance	Dominance is the state of having high social status relative to other individuals, who react submissively to dominant individuals.
Substrate	A substrate is a molecule which is acted upon by an enzyme. Each enzyme recognizes only the specific substrate of the reaction it catalyzes. A surface in or on which an organism lives.
Abiotic	Abiotic refers to the nonliving portion of an ecosystem. It includes soil, rock, water, and the atmosphere.
Infauna	Infauna are those organisms that live within particulate media such as sediments or soil. They are most common in the subtidal and deeper zones.
Sulfide	One of the minerals that is abundant in the hot water that seeps through hydrothermal vents is sulfide.
Salt marsh	Salt marsh refers to a grassy area that extends along the shores of estuaries and sheltered coasts in temperate and subpolar regions.
Seaweed	Phycologists consider seaweed to refer any of a large number of marine benthic algae that are multicellular, macrothallic (large-bodied), and thus differentiated from most algae that tend to be microscopic in size.
Genus	In biology, a genus is a taxonomic grouping. That is, in the classification of living organisms, a genus is considered to be distinct from other such genera. A genus has one or more species: if it has more than one species these are likely to be morphologically more similar than species belonging to different genera.
Hydrogen	Hydrogen is a chemical element in the periodic table that has the symbol H and atomic number 1. At standard temperature and pressure it is a colorless, odorless, nonmetallic, univalent, tasteless, highly flammable diatomic gas.
Hydrogen sulfide	The gas that is produced in anoxic sediments is referred to as hydrogen sulfide.
Crustacean	Member of a group of marine arthropods that contains among others shrimps, crabs, crayfish, and lobsters is called crustacean.

Bivalve	Bivalve refers to a member of a group of mollusks that includes clams, mussels, scallops, and oysters. They typically have two-part shells, with both parts being more or less symmetrical.
Generation time	Generation time refers to the time between the birth of a parent and the birth of its offspring. The time interval between successive cell divisions.
Residue	A residue refers to a portion of a larger molecule, a specific monomer of a polysaccharide, protein or nucleic acid.
Detoxification	Detoxification in general is the removal of toxic substances from the body. It is one of the functions of the liver and kidneys, but can also be achieved artificially by techniques such as dialysis and (in a very limited number of cases) chelation therapy.
Siphon	The tube-like extension through which water flows in and out of the mantle cavity in bivalves and cephalopods, and in tunicates is called siphon.
Biology	Biology is the branch of science dealing with the study of life. It is concerned with the characteristics, classification, and behaviors of organisms, how species come into existence, and the interactions they have with each other and with the environment.
Benthos	Benthos are the organisms and habitats of the sea floor; in freshwater biology they are the organisms and habitats of the bottoms of lakes, rivers, and creeks.
Runner	A horizontally growing stem that may develop new plants at nodes that touch the soil is called runner.
Meiofauna	Meiofauna are small benthic invertebrates that live in both marine and fresh water environments. The term Meiofauna loosely defines a group of organisms by their size, larger than Microfauna but smaller than Macrofauna, rather than a taxonomic grouping.
Biomass	Biomass is organic non-fossil material, collectively. In other words, biomass comprises the mass of all biological organisms, dead or alive, excluding biological mass that has been transformed by geological processes into substances such as coal or petroleum.
Abyssal plain	Abyssal plain is a flat or very gently sloping area of the deep ocean basin floor. They are among the Earth's flattest and smoothest regions and the least explored. They cover approximately 40% of the ocean floor and reach depths between 2,200 and 5,500 m). They generally lie between the foot of a continental rise and a mid-oceanic ridge.
Standing stock	Standing stock refers to the total amount, or biomass, of an organism at a given time.
Continental slope	Continental slope the part of the continental margin that lies between the continental shelf and the bottom of the ocean. Sunlight does not penetrate this area, and mostly it is home to scavengers. It is characterized by a relatively steep slope of 3° to 6°.
Sargasso sea	Sargasso sea refers to the region of the Atlantic Ocean north of the West Indies that is characterized by floating masses of Sargasso weed, a brown alga.
Primary production	Primary production is the production of organic compounds from inorganic materials principally through the process of photosynthesis.
Relative abundance	Differences in the abundance of species within a community is relative abundance.
Morphology	The scientific study of organic form, including both its development and function is morphology.
Muscle	Muscle is a contractile form of tissue. It is one of the four major tissue types, the other three being epithelium, connective tissue and nervous tissue. Muscle contraction is used to move parts of the body, as well as to move substances within the body.
Metabolic rate	Energy expended by the body per unit time is called metabolic rate.

Stomach	The stomach is an organ in the alimentary canal used to digest food. It's primary function is not the absorption of nutrients from digested food; rather, the main job of the stomach is to break down large food molecules into smaller ones, so that they can be absorbed into the blood more easily.
Protein	A protein is a complex, high-molecular-weight organic compound that consists of amino acids joined by peptide bonds. They are essential to the structure and function of all living cells and viruses. Many are enzymes or subunits of enzymes.
Starch	Biochemically, starch is a combination of two polymeric carbohydrates (polysaccharides) called amylose and amylopectin.
Electrode	An electrode is a conductor used to make contact with a nonmetallic part of a circuit (e.g. a semiconductor, an electrolyte or a vacuum).
Radioactive	A term used to describe the property of releasing energy or particles from an unstable atom is called radioactive.
Carbon	Carbon is a chemical element in the periodic table that has the symbol C and atomic number 6. An abundant nonmetallic, tetravalent element, carbon has several allotropic forms.
Metabolism	Metabolism is the biochemical modification of chemical compounds in living organisms and cells. This includes the biosynthesis of complex organic molecules (anabolism) and their breakdown (catabolism).
Atmosphere	Earth's atmosphere is a layer of gases surrounding the planet Earth and retained by the Earth's gravity. It contains roughly 78% nitrogen and 21% oxygen, with trace amounts of other gases.
Dissolved oxygen	Oxygen gas molecules dissolved in water. Fish and other aquatic organisms are dependent on dissolved oxygen for respiration. Therefore concentration of dissolved oxygen is a measure of water quality.
Limiting factor	In biology, agricultural science, physiology, and ecology, a limiting factor is one that controls a process, such as organism growth or species population size or distribution. The availability of food, predation pressure, or availability of shelter are examples of factors that could be limiting for a species population in a specific area.
Runoff	Water entering rivers, lakes, reservoirs, or the ocean from land surfaces is runoff.
Biodiversity	Biodiversity is the totality of genes, species, and ecosystems of a region.
Species diversity	Species diversity refers to a weighted representation of the species of organisms living in a region; large and common species are given greater weight than are small and rare ones.
Rarefaction	A method of determining the relationship between species diversity and sample size by randomly deleting individuals from a sample is rarefaction.
Biota	Biota is the plant and animal life of a region or area or geological period. The biota of the Earth is known as the biosphere.
Extinction	In biology and ecology, extinction is the ceasing of existence of a species or group of taxa. The moment of extinction is generally considered to be the death of the last individual of that species.The death of all members of a species is extinction.
Gastropod	A member of the largest group of mollusks, including snails and slugs is a gastropod.
Snail	The name snail applies to most members of the molluscan class Gastropoda that have coiled shells. Other gastropods, which lack a conspicuous shell, are commonly called slugs, and are scattered throughout groups that primarily include snails.
Isopod	Isopod is one of the most diverse orders of Crustaceans, with many species living in all

	environments, and are common in shallow marine waters. Unlike most crustaceans, isopods are successful on land (suborder Oniscidea, woodlice), although their greatest diversity remains in the deep sea (suborder Asellota).
Isopoda	Isopoda is one of the most diverse orders of Crustaceans, with many species living in all environments, and are common in shallow marine waters. Unlike most crustaceans, they are successful on land although their greatest diversity remains in the deep sea.
Pulse	The rhythmic stretching of the arteries caused by the pressure of blood forced through the arteries by contractions of the ventricles during systole is a pulse.
Interspecific competition	Competition between populations of two or more species for one or more of the same limited resources of food, sunlight, water, soil, nutrients, or space. Interspecific competition may inhibit population growth and help structure communities.
Speciation	Speciation refers to the evolutionary process by which new biological species arise. All forms of speciation have actually taken place over the course of evolution, though it still remains a subject of debate as to the relative importance of each mechanism in driving biodiversity.
Biotic	Biotic refers to life or living organisms.
Chitin	Chitin refers to strong but flexible nitrogenous polysaccharide found in fungal cells and in the exoskeleton of arthropods.
Phylum	Phylum is the second highest taxonomic classification for the kingdom Animalia (animals), between kingdom level and class level.
Oxidation	Oxidation refers to the loss of electrons from a substance involved in a redox reaction; always accompanies reduction.
Mollusk	The mollusk is the large and diverse phylum Mollusca, which includes a variety of familiar creatures well-known for their decorative shells or as seafood. These range from tiny snails, clams, and abalone to the octopus, cuttlefish and squid (which are considered the most intelligent invertebrates).
Sulfur	Sulfur is the chemical element in the periodic table that has the symbol S and atomic number 16. It is an abundant, tasteless, odorless, multivalent non-metal. Sulfur, in its native form, is a yellow crystaline solid. In nature, it can be found as the pure element or as sulfide and sulfate minerals.
Symbiosis	Symbiosis is an interaction between two organisms living together in more or less intimate association or even the merging of two dissimilar organisms. The term host is usually used for the larger (macro) of the two members of a symbiosis.
Parasite	A parasite is an organism that spends a significant portion of its life in or on the living tissue of a host organism and which causes harm to the host without immediately killing it. They also commonly show highly specialized adaptations allowing them to exploit host resources.
Hemoglobin	Hemoglobin is the iron-containing oxygen-transport metalloprotein in the red cells of the blood in mammals and other animals. Hemoglobin transports oxygen from the lungs to the rest of the body, such as to the muscles, where it releases the oxygen load.
Host	Host is an organism that harbors a parasite, mutual partner, or commensal partner; or a cell infected by a virus.
Amino acid	An amino acid is any molecule that contains both amino and carboxylic acid functional groups. They are the basic structural building units of proteins. They form short polymer chains called peptides or polypeptides which in turn form structures called proteins.

Cysteine	Cysteine is a naturally occurring hydrophobic amino acid which has a thiol group and is found in most proteins, though only in small quantities.
Digestive tract	The digestive tract is the system of organs within multicellular animals which takes in food, digests it to extract energy and nutrients, and expels the remaining waste.
Gill	An extension of the body surface of an aquatic animal, specialized for gas exchange and/or suspension feeding is called a gill.
Hydrocarbon	A chemical compound composed only of the elements carbon and hydrogen is called hydrocarbon.
Sugar	A sugar is the simplest molecule that can be identified as a carbohydrate. These include monosaccharides and disaccharides, trisaccharides and the oligosaccharides. The term "glyco-" indicates the presence of a sugar in an otherwise non-carbohydrate substance.
Archaebacteria	Archaebacteria refers to a group of bacteria that are among the most primitive still in existence, characterized by the absence of peptidogylcan in their cell walls, a feature that distinguishes them from all other bacteria.
Limpet	Limpet is a marine mollusk in the family Acmaeidae with flattened, cone-shaped shells. They live throughout the intertidal zone, from the high zone (upper littoral) to the shallow subtidal on the rocky coasts of most oceans. They can be commonly found attached to rocks, looking like little disks or bumps on the rock surface.
Crab	The term crab is often applied to several different groups of short (nose to tail) decapod crustaceans with thick exoskeletons, but only members of the Brachyura are true.
Symbiont	The smaller participant in a symbiotic relationship, living in or on the host is a symbiont.
Fauna	Fauna (animals), a collective term for animal life, as distinct from flora (plant life).
Symbiotic nitrogen-fixing bacteria	Bacteria that live in the roots of certain kinds of plants, where they convert nitrogen gas molecules into compounds that plants can use is called symbiotic nitrogen-fixing bacteria.
Nitrogen	A colorless and tasteless and mostly inert diatomic non-metal gas that is an essential constituent of proteins is nitrogen.
Biochemistry	Biochemistry studies how complex chemical reactions give rise to life. It is a hybrid branch of chemistry which specialises in the chemical processes in living organisms.
Sunlight	Sunlight in the broad sense is the total spectrum of electromagnetic radiation given off by the Sun.
Enzyme	An enzyme is a protein that catalyzes, or speeds up, a chemical reaction. They are essential to sustain life because most chemical reactions in biological cells would occur too slowly, or would lead to different products, without them.
Cell	The cell is the structural and functional unit of all living organisms, and is sometimes called the "building block of life."
Molecule	A molecule is the smallest particle of a pure chemical substance that still retains its chemical composition and properties.
Cofactor	A cofactor is any substance that needs to be present in addition to an enzyme to catalyze a certain reaction.
Cell membrane	A component of every biological cell, the selectively permeable cell membrane is a thin and structured bilayer of phospholipid and protein molecules that envelopes the cell. It separates a cell's interior from its surroundings and controls what moves in and out.
Dehydrogenase	An enzyme that catalyzes a chemical reaction during which one or more hydrogen atoms are

	removed from a molecule is dehydrogenase.
Adaptation	A biological adaptation is an anatomical structure, physiological process or behavioral trait of an organism that has evolved over a period of time by the process of natural selection such that it increases the expected long-term reproductive success of the organism.
Trophic structure	The different feeding relationships in an ecosystem, which determine the route of energy flow and the pattern of chemical cycling is called trophic structure.
Hydrothermal vent	A hydrothermal vent is a fissure in a planet's surface from which geothermally heated water issues. Hydrothermal vents are commonly found in places that are also volcanically active, where hot magma is relatively near the planet's surface.
Blood	Blood is a circulating tissue composed of fluid plasma and cells. The main function of blood is to supply nutrients (oxygen, glucose) and constitutional elements to tissues and to remove waste products.
Orbital	Orbital refers to a volume of space in which a given electron occurs 90% of the time. It is the quantum state of the individual electrons in the electron cloud around a single atom.
Oligotrophic	Oligotrophic refers to any environment which offers little to sustain life. This term is usually used to describe bodies of water with very low nutrient levels.
Ecology	Ecology is the scientific study of the distribution and abundance of living organisms and how these properties are affected by interactions between the organisms and their environment.
Benthic community	The benthic community is an assemblage of interacting organisms, including those forming structural elements, found at or near the bottom of a body of water.
Species richness	Species richness is perhaps the simplest measure of biodiversity. The larger the number, the more species in an area. It is referred to in equations as S. There is a strong inverse correlation in many groups between species richness and latitude - the further from the equator you go, the fewer species you will find.
Microbiology	Microbiology is the study of microorganisms, including viruses, viroids, prions, prokaryotes and simple eukaryotes (e.g. molds and fungi).
Systematics	In biology, systematics is the study of the diversity of organism characteristics, and especially how they relate evolutionarily. The term "systematics" may or may not overlap with "taxonomy" which concerns itself with scientific classification of species and other taxa.
Nekton	Animals, such as fish, that can swim against currents of water are referred to as nekton.
Bioaccumulation	Bioaccumulation means to accumulate in a biological system. However, it is commonly taken to measure the uptake over time of a substance, that can accumulate in a biological system.
Evolution	In biology, evolution is the process by which novel traits arise in populations and are passed on from generation to generation. Its action over large stretches of time explains the origin of new species and ultimately the vast diversity of the biological world.

Speciation	Speciation refers to the evolutionary process by which new biological species arise. All forms of speciation have actually taken place over the course of evolution, though it still remains a subject of debate as to the relative importance of each mechanism in driving biodiversity.
Extinction	In biology and ecology, extinction is the ceasing of existence of a species or group of taxa. The moment of extinction is generally considered to be the death of the last individual of that species.The death of all members of a species is extinction.
Species	Group of similarly constructed organisms capable of interbreeding and producing fertile offspring is a species.
Origin of Species	First published in 1859, The Origin of Species by British naturalist Charles Darwin is one of the pivotal works in scientific history, and arguably the pre-eminent work in biology.
Climate	Weather condition of an area including especially prevailing temperature and average daily/yearly rainfall over a long period of time is called climate.
Biota	Biota is the plant and animal life of a region or area or geological period. The biota of the Earth is known as the biosphere.
Predation	Interaction in which one organism uses another, called the prey, as a food source is referred to as predation.
Population	Group of organisms of the same species occupying a certain area and sharing a common gene pool is referred to as population.
Geographic isolation	A condition in which part of the gene pool is separated by geographic barriers from the rest of the population is referred to as geographic isolation.
Natural selection	Natural selection is the process by which biological individuals that are endowed with favorable or deleterious traits end up reproducing more or less than other individuals that do not possess such traits.
Gene flow	The gain or loss of alleles from a population by the movement of individuals or gametes into or out of the population is called gene flow.
Gene	Gene refers to a discrete unit of hereditary information consisting of a specific nucleotide sequence in DNA . Most of the genes of a eukaryote are located in its chromosomal DNA; a few are carried by the DNA of mitochondria and chloroplasts.
Isthmus	An isthmus is a narrow strip of land that is bordered on two sides by water and connects two larger land masses. It is the inverse of a strait (which lies between two land masses and connects two larger bodies of water).
Enzyme	An enzyme is a protein that catalyzes, or speeds up, a chemical reaction. They are essential to sustain life because most chemical reactions in biological cells would occur too slowly, or would lead to different products, without them.
Interspecific competition	Competition between populations of two or more species for one or more of the same limited resources of food, sunlight, water, soil, nutrients, or space. Interspecific competition may inhibit population growth and help structure communities.
Geographic barriers	Geographic features that keep different portions of a species from exchanging genes are called geographic barriers.
Habitat	Habitat refers to a place where an organism lives; an environmental situation in which an organism lives.
Geographic range	The geographic area in which a population lives is a geographic range.
Lagoon	A lagoon is a body of comparatively shallow salt water separated from the deeper sea by a

	shallow or exposed sandbank, coral reef, or similar feature. Thus, the enclosed body of water behind a barrier reef or barrier islands or enclosed by an atoll reef is called a lagoon.
Biological interaction	A biological interaction results from the fact that organisms in an ecosystem interact with each other, in the natural world, no organism is an autonomous entity isolated from its surroundings.
Dominant species	Dominant species refers to those species in a community that have the highest abundance or highest biomass. These species exert a powerful control over the occurrence and distribution of other species.
Predator	A predator is an animal or other organism that hunts and kills other organisms for food in an act called predation.
Dispersal	Biological dispersal refers to those processes by which a species maintains or expands the distribution of a population. Dispersal implies movement—movement away from an existing population or away from the parent organisms.
Snail	The name snail applies to most members of the molluscan class Gastropoda that have coiled shells. Other gastropods, which lack a conspicuous shell, are commonly called slugs, and are scattered throughout groups that primarily include snails.
Salinity	Salinity refers to the total amount of salts dissolved in seawater. It is generally expressed in parts per thousand.
Gradient	Gradient refers to a difference in concentration, pressure, or electrical charge between two regions.
Seafloor spreading	Seafloor spreading is a part of the theory of plate tectonics. Seafloor spreading is the process by which continental drift occurs. This mechanism, which is a more accurate version of Alfred Wegener's original drift of continents that "plow" through the sea.
Continental drift	Continental drift refers to movement of continents with respect to one another over the earth's surface.
Plankton	Plankton are drifting organisms that inhabit the water column of oceans, seas, and bodies of fresh water. Plankton abundance and distribution are strongly dependent on factors such as ambient nutrients concentrations, the physical state of the water column, and the abundance of other plankton.
Benthos	Benthos are the organisms and habitats of the sea floor; in freshwater biology they are the organisms and habitats of the bottoms of lakes, rivers, and creeks.
Biogeographic region	A continental-scale part of Earth that has a biota distinct from that of other such regions is called biogeographic region.
Conception	Conception is fusion of gametes to form a new organism. In animals, the process involves a sperm fusing with an ovum, which eventually leads to the development of an embryo.
Continental shelf	The shallow part of the seafloor immediately adjacent to a continent is called the continental shelf.
Species diversity	Species diversity refers to a weighted representation of the species of organisms living in a region; large and common species are given greater weight than are small and rare ones.
Microhabitat	A microhabitat is the immediate surroundings and other physical factors of an individual plant or animal within its habitat.
Coral reef	A coral reef is a type of biotic reef that develops in tropical waters by the growth of coralline algae, hermatypic corals, and other marine organisms.
Mitochondrial	Mitochondrial DNA is DNA not located in the nucleus of the cell but in the mitochondria,

DNA	parts of the cell that generate fuel in the form of adenosine triphosphate, which drives the varied machinery of the cell. Unlike most of the cell, the function of which is defined by the nuclear DNA, the mitochondria have their own DNA and are assumed to have evolved separately.
Distribution	Distribution in pharmacology is a branch of pharmacokinetics describing reversible transfer of drug from one location to another within the body.
Monophyletic	A group is monophyletic if it consists of a common ancestor and all its descendants. A taxonomic group that contain organisms but not their common ancestor is called polyphyletic, and a group that contains some but not all descendants of the most recent common ancestor is called paraphyletic.
Bivalve	Bivalve refers to a member of a group of mollusks that includes clams, mussels, scallops, and oysters. They typically have two-part shells, with both parts being more or less symmetrical.
Mollusk	The mollusk is the large and diverse phylum Mollusca, which includes a variety of familiar creatures well-known for their decorative shells or as seafood. These range from tiny snails, clams, and abalone to the octopus, cuttlefish and squid (which are considered the most intelligent invertebrates).
Taxon	A taxon, is a grouping of organisms (named or unnamed). Once named, a taxon will usually have a rank and can be placed at a particular level in a hierarchy.
Glaciation	A glaciation, often called an ice age, is a geological phenomenon in which massive ice sheets form in the Arctic and Antarctic and advance toward the equator.
Species richness	Species richness is perhaps the simplest measure of biodiversity. The larger the number, the more species in an area. It is referred to in equations as S. There is a strong inverse correlation in many groups between species richness and latitude - the further from the equator you go, the fewer species you will find.
Gastropod	A member of the largest group of mollusks, including snails and slugs is a gastropod.
Deep sea	The deep sea refers to those areas of oceans to which little or no light penetrates.
Annelid	Annelid refers to member of a phylum of invertebrates that contains segmented worms, such as the earthworm and the clam worm.
Coral	Group of cnidarians having a calcium carbonate skeleton that participate in the formation of reefs is coral.
Water column	The vertical column of seawater that extends from the surface to the bottom is called water column.
Abyssal plain	Abyssal plain is a flat or very gently sloping area of the deep ocean basin floor. They are among the Earth's flattest and smoothest regions and the least explored. They cover approximately 40% of the ocean floor and reach depths between 2,200 and 5,500 m). They generally lie between the foot of a continental rise and a mid-oceanic ridge.
Continental rise	The gently sloping area at the base of the continental slope is called the continental rise.
Regional diversity	The number of species or other taxa occurring within a large geographic area encompassing many biomes or habitats is called regional diversity.
Trait	In biology, a trait or character is a genetically inherited feature of an organism: Eye color is a character; brown eyes and blue eyes is a trait.
Fauna	Fauna (animals), a collective term for animal life, as distinct from flora (plant life).
Vicariance	The separation or division of a group of organisms by a geographic barrier, such as a mountain or a body of water, resulting in differentiation of the original group into new

	varieties or species is called vicariance.
Oligotrophic	Oligotrophic refers to any environment which offers little to sustain life. This term is usually used to describe bodies of water with very low nutrient levels.
Detritus	Detritus is organic waste material from decomposing dead plants or animals.
Fossil record	The chronicle of evolution over millions of years of geologic time engraved in the order in which fossils appear in rock strata is called a fossil record.
Crinoids	Crinoids refers to the echinoderms with a small, cup-shaped body and feathery arms: the sea lilies and feather stars .
Fossil	A preserved remnant or impression of an organism that lived in the past is referred to as fossil.
Phanerozoic	The Phanerozoic Eon is the period of geologic time during which abundant animal life has existed. It covers roughly 545 million years and goes back to the time when diverse hard-shelled animals first appeared.
Crust	A crust is the outer layer of a planet, part of its lithosphere. Planetary crust is generally composed of a less dense material than that of its deeper layers. The crust of the Earth is composed mainly of basalt and granite.
Carboniferous	The Carboniferous is a major division of the geologic timescale that extends from the end of the Devonian period, about 359.2 Ma (million years ago), to the beginning of the Permian period, about 299.0 Ma (ICS 2004).
Radioactive	A term used to describe the property of releasing energy or particles from an unstable atom is called radioactive.
Ordovician	The Ordovician period is the second of the six (seven in North America) periods of the Paleozoic era. It follows the Cambrian period and is followed by the Silurian period.
Paleozoic	The Paleozoic Era is a major division of the geologic timescale, one of four geologic eras. The division of time into eras, the largest division of geologic time, dates back to Giovanni Arduino in the 18th century, although his original name for the era now called the Paleozoic was "Primitive".
Triassic	The Triassic is a geologic period that extends from about 245 to 202 Ma (million years ago). As the first period of the Mesozoic Era, the Triassic follows the Permian and is followed by the Jurassic.
Devonian	During the Devonian Period the first fish evolved legs and started to walk on land as amphibians, and the first arthropods like insects and spiders also started to colonize terrestrial habitats.
Jurassic	The Jurassic period is a major unit of the geologic timescale that extends from about 200 Ma (million years ago) at the end of the Triassic to 146 Ma at the beginning of the Cretaceous.
Cambrian	The Cambrian is a major division of the geologic timescale that begins about 542 Ma (million years ago) at the end of the Proterozoic eon and ended about 488.3 Ma with the beginning of the Ordovician period (ICS, 2004).
Isotope	An isotope is a form of an element whose nuclei have the same atomic number - the number of protons in the nucleus - but different mass numbers because they contain different numbers of neutrons.
Radiometric dating	Radiometric dating is a technique used to date materials based on a knowledge of the decay rates of naturally occurring isotopes, and the current abundances.
Mass extinction	Mass extinction refers to abrupt disappearance of a large fraction of a biota, thought to be

caused by such environmental catastrophes as meteor impacts; significant mass extinctions occurred at the end of the Permian and Cretaceous periods.

Cretaceous — The Cretaceous period is one of the major divisions of the geologic timescale, reaching from the end of the Jurassic period, about 146 million years ago (Ma), to the beginning of the Paleocene epoch of the Tertiary period (65.5 Ma).

Dinosaur — The dinosaur was a vertebrate animal that dominated the terrestrial ecosystem for over 160 million years, first appearing approximately 230 million years ago. At the end of the Cretaceous period 65 million years ago, they suffered a catastrophic extinction, which ended their dominance on land.

Permian — The Permian is a geologic period that extends from about 299.0 Ma to 248.0 Ma. It is the last period of the Palaeozoic Era. As with most older geologic periods, the strata that define the Permian are well identified, but the exact date of the period's start is uncertain by a few million years.

Biodiversity — Biodiversity is the totality of genes, species, and ecosystems of a region.

Invertebrate — Invertebrate is a term coined by Jean-Baptiste Lamarck to describe any animal without a spinal column. It therefore includes all animals except vertebrates (fish, reptiles, amphibians, birds and mammals).

Seaweed — Phycologists consider seaweed to refer any of a large number of marine benthic algae that are multicellular, macrothallic (large-bodied), and thus differentiated from most algae that tend to be microscopic in size.

Polymorphism — The presence in a population of more than one allele of a gene at a frequency greater than that of newly arising mutations is referred to as polymorphism.

Sequencing — Determining the order of nucleotides in a DNA or RNA molecule or the order of amino acids in a protein is sequencing.

Ecosystem — In general terms an ecosystem can be thought of as an assemblage of organisms (plant, animal and other living organisms living together with their environment, functioning as a loose unit. That is, a dynamic and complex whole, interacting as an "ecological unit".

Mussel — A mussel is a bivalve mollusk that can be found in lakes, rivers, creeks, intertidal areas, and throughout the ocean. The saltwater mussels (family Mytilidae) and freshwater mussels (family Unionidae) are not thought to be closely related and are grouped in different subclasses, despite considerable similarities in appearance.

Algae — The algae consist of several different groups of living organisms that capture light energy through photosynthesis, converting inorganic substances into simple sugars with the captured energy.

Community ecology — The study of how interactions between species affect community structure and organization is referred to as community ecology.

Carnivore — An animal that eats a diet consisting solely of meat is referred to as a carnivore.

Food web — A network of interconnecting food chains is referred to as a food web.

Fragmentation — Fragmentation is a form of asexual reproduction where an organism is split into fragments. The splitting may or may not be intentional. Each of these fragments develops into a mature, fully grown new organism, and if the organism is split any further the process is repeated.

Erosion — Erosion is the displacement of solids (soil, mud, rock, and other particles) by the agents of wind, water, ice, movement in response to gravity, or living organisms.

Spawning — A method of external fertilization in which male and female parents shed gametes into the

	water, and sperm must swim through the water to reach the eggs is referred to as spawning.
Colonization	Colonization is the act where life forms move into a distant area where their kind is sparse or not yet existing at all and set up new settlements in the area. Colonization applies to all life forms in a sense though it is most often used in reference to insects and humans.
Residue	A residue refers to a portion of a larger molecule, a specific monomer of a polysaccharide, protein or nucleic acid.
Stem	Stem refers to that part of a plant's shoot system that supports the leaves and reproductive structures.
Exotic species	Exotic species refers to a foreign species introduced into an ecosystem where it did not evolve; such species may florish and outcompete native species.
Zooplankton	Nonphotosynthetic protists that feed on other plankton and are abundant in marine and freshwater environments are zooplankton.
Ctenophore	A ctenophore is a jellyfish-like animal commonly called a comb jelly. Despite their superficial similarities with jellyfish, they have a much more complex anatomy.
Vector	A vector is an organism that does not cause disease itself but which spreads infection by conveying pathogens from one host to another.
Larva	A free-living, sexually immature form in some animal life cycles that may differ from the adult in morphology, nutrition, and habitat is called larva.
Phytoplankton	Phytoplankton refers to the autotrophic component of the plankton that drifts in the water column. Phytoplankton obtain energy through a process called photosynthesis, and so must live in the well-lit surface layer (termed the euphotic zone) of an ocean, sea, or lake.
Protozoa	Protozoa are single-celled eukaryotes (organisms with nuclei) that show some characteristics usually associated with animals, most notably mobility and heterotrophy. They are often grouped in the kingdom Protista together with the plant-like algae.
Bacteria	The domain that contains procaryotic cells with primarily diacyl glycerol diesters in their membranes and with bacterial rRNA. Bacteria also is a general term for organisms that are composed of procaryotic cells and are not multicellular.
Trade-off	Trade-off refers to a balancing of factors, all of which are not attainable at the same time, especially traits that tend to increase the evolutionary fitness or performance of the individual.
Migration	Migration occurs when living things move from one biome to another. In most cases organisms migrate to avoid local shortages of food, usually caused by winter. Animals may also migrate to a certain location to breed, as is the case with some fish.
Estuary	An estuary is a semi-enclosed coastal body of water which has a free connection with the open sea and within which sea water mixes with fresh water. The key feature of an estuary is that it is a mixing place for sea water and a stream or river to supply fresh water.
Crab	The term crab is often applied to several different groups of short (nose to tail) decapod crustaceans with thick exoskeletons, but only members of the Brachyura are true.
Grazing	Grazing refers to consumption of a portion of a plant's tissues, generallyapplied to grasses and other herbaceous vegetation; removing plant or algal cover from a substrate, generally applied to aquatic systems.
Egg	An egg is the zygote, resulting from fertilization of the ovum. It nourishes and protects the embryo.
Endemic species	In biology and ecology endemic species means exclusively native to a place or biota, in

	contrast to cosmopolitan or introduced.
Reptile	Member of a class of terrestrial vertebrates with internal fertilization, scaly skin, and an egg with a leathery shell is called reptile.
Source population	A population that can provide members to other populations of the species because it lives in a favorable area, and the birthrate is most likely higher than the death rate is called a source population.
Molecular marker	A segment of DNA found at a specific site in a genome that has variants which can be recognized and followed is called a molecular marker.
Genetic marker	A genetic marker is a specific piece of DNA with a known position on the genome. It is a genetic technique to follow a certain disease or gene. It is known that pieces of DNA that lie near each other, tend to be inherited to the next organism together.
Copepod	The group of small crustaceans that is an important member of marine and freshwater plankton communities is copepod. They are very important food organisms for small fish, whales, and other crustaceans such as krill in the ocean and in fresh water.
Adaptation	A biological adaptation is an anatomical structure, physiological process or behavioral trait of an organism that has evolved over a period of time by the process of natural selection such that it increases the expected long-term reproductive success of the organism.
Lamprey	A lamprey is a jawless fish with a toothed, funnel-like sucking mouth, with which most species bore into the flesh of other fishes to suck their blood.
Substratum	Substratum is the underlying layer, or base to which a lichen is fixed.
Invasive species	The term invasive species refers to a subset of those species defined as introduced species or non-indigenous species. Invasive species can alter ecological relationships among native species and can affect ecosystem function, economic value of ecosystems, and human health.
Prediction	Step of the scientific method which follows the formulation of a hypothesis and assists in creating the experimental design is referred to as prediction.
Population density	The number of organisms of a species per unit area is called population density.
Conservation genetics	Conservation genetics is a science that aims to apply genetic methods to deal with the maintenance, loss, and restoration of biodiversity.
Sessile	In botany, sessile means without a stalk, as in flowers or leaves that grow directly from the stem. In zoology, sessile organisms are those which are not able to move about. They are usually permanently attached to a solid substrate of some kind, such as a rock, or the hull of a ship in the case of barnacles.
DNA sequencing	The process of determining the chemical composition of a DNA molecule is called DNA sequencing.
Endangered species	An endangered species is a population of organisms (frequently but not always a taxonomic species) which is either (a) so few in number or (b) threatened by changing environmental or predation parameters that it is at risk of becoming extinct.
Population structure	Population structure refers to the proportions of individuals in a population belonging to different age classes. Also, the distribution of the population in space.
Orbital	Orbital refers to a volume of space in which a given electron occurs 90% of the time. It is the quantum state of the individual electrons in the electron cloud around a single atom.
Protein	A protein is a complex, high-molecular-weight organic compound that consists of amino acids joined by peptide bonds. They are essential to the structure and function of all living cells

	and viruses. Many are enzymes or subunits of enzymes.
Biogeography	Biogeography is the science which deals with patterns of species distribution and the processes that result in such patterns.
Evolution	In biology, evolution is the process by which novel traits arise in populations and are passed on from generation to generation. Its action over large stretches of time explains the origin of new species and ultimately the vast diversity of the biological world.
Seagrass	Seagrass are flowering plants from four plant families (Posidoniaceae, Zosteraceae, Hydrocharitaceae and Cymodoceaceae) that grow in the marine saline environment.
Genetic structure	The frequencies of alleles and genotypes in a population is a genetic structure.
Sea urchin	A sea urchin is a spiny sea creature of the class Echinoidea found in oceans all over the world. Their shell, which biologists call the test, is globular in shape, and covered with spines.
Systematics	In biology, systematics is the study of the diversity of organism characteristics, and especially how they relate evolutionarily. The term "systematics" may or may not overlap with "taxonomy" which concerns itself with scientific classification of species and other taxa.
Ecology	Ecology is the scientific study of the distribution and abundance of living organisms and how these properties are affected by interactions between the organisms and their environment.
Crustacea	The animal phylum that includes mostly aquatic animals such as crabs, lobsters, crayfish, and shrimp is referred to as crustacea.
Decapoda	The decapoda are an order of crustaceans within the class Malacostraca, including many familiar groups, such as crabs, lobsters, prawns and shrimps, but also some families that are less well known.
Cnidaria	Cnidaria is a phylum containing some 10,000 species of relatively simple animals found exclusively in aquatic environments (most species are marine). They get their name from cnidocytes, which are specialized cells that carry stinging organelles.
Introduced species	Introduced species refers to a species that humans move from the species' native location to a new geographic region; sometimes called an exotic species.
Biology	Biology is the branch of science dealing with the study of life. It is concerned with the characteristics, classification, and behaviors of organisms, how species come into existence, and the interactions they have with each other and with the environment.
Trophic structure	The different feeding relationships in an ecosystem, which determine the route of energy flow and the pattern of chemical cycling is called trophic structure.

Bark	Bark is the outermost layer of stems and roots of woody plants such as trees. It overlays the wood and consists of three layers, the cork, the phloem, and the vascular cambium.
Population	Group of organisms of the same species occupying a certain area and sharing a common gene pool is referred to as population.
Extinction	In biology and ecology, extinction is the ceasing of existence of a species or group of taxa. The moment of extinction is generally considered to be the death of the last individual of that species.The death of all members of a species is extinction.
Species	Group of similarly constructed organisms capable of interbreeding and producing fertile offspring is a species.
Protein	A protein is a complex, high-molecular-weight organic compound that consists of amino acids joined by peptide bonds. They are essential to the structure and function of all living cells and viruses. Many are enzymes or subunits of enzymes.
International Whaling Commission	An agency that regulates whaling around the world--the International Whaling Commission .
Renewable resource	A renewable resource is any natural resource that is depleted at a rate slower than the rate at which it regenerates.
Migration	Migration occurs when living things move from one biome to another. In most cases organisms migrate to avoid local shortages of food, usually caused by winter. Animals may also migrate to a certain location to breed, as is the case with some fish.
Fish migration	Fish migration occurs on a regular basis, on time scales ranging from daily to annual, and with distances ranging from a few meters to thousands of kilometers. The purpose usually relates to either feeding or breeding; in some cases the reason is still unknown.
Spawning	A method of external fertilization in which male and female parents shed gametes into the water, and sperm must swim through the water to reach the eggs is referred to as spawning.
Geographic range	The geographic area in which a population lives is a geographic range.
Molecular marker	A segment of DNA found at a specific site in a genome that has variants which can be recognized and followed is called a molecular marker.
Natural selection	Natural selection is the process by which biological individuals that are endowed with favorable or deleterious traits end up reproducing more or less than other individuals that do not possess such traits.
Genetic marker	A genetic marker is a specific piece of DNA with a known position on the genome. It is a genetic technique to follow a certain disease or gene. It is known that pieces of DNA that lie near each other, tend to be inherited to the next organism together.
DNA markers	DNA markers refers to a purified fragment of DNA labeled with a radioactive isotope or fluorescent dye and used to identify complementary sequences by means of hybridization.
Mitochondrial DNA	Mitochondrial DNA is DNA not located in the nucleus of the cell but in the mitochondria, parts of the cell that generate fuel in the form of adenosine triphosphate, which drives the varied machinery of the cell. Unlike most of the cell, the function of which is defined by the nuclear DNA, the mitochondria have their own DNA and are assumed to have evolved separately.
Radioactive	A term used to describe the property of releasing energy or particles from an unstable atom is called radioactive.
Carbon	Carbon is a chemical element in the periodic table that has the symbol C and atomic number 6.

	An abundant nonmetallic, tetravalent element, carbon has several allotropic forms.
Salt	Salt is a term used for ionic compounds composed of positively charged cations and negatively charged anions, so that the product is neutral and without a net charge.
Amplification	In molecular biology, amplification (polymerase chain reaction or PCR) is the method for creating multiple copies of DNA (or RNA) without using a living organism, such as E. coli or yeast.
Polymerase	DNA or RNA enzymes that catalyze the synthesis of nucleic acids on preexisting nucleic acid templates, assembling RNA from ribonucleotides or DNA from deoxyribonucleotides is referred to as polymerase.
Sperm Whale	The Sperm Whale (Physeter macrocephalus) is the largest of all toothed whales and is believed to be the largest toothed animal to ever inhabit Earth.
Hybrid	Hybrid refers to the offspring of parents of two different species or of two different varieties of one species; the offspring of two parents that differ in one or more inherited traits; an individual that is heterozygous for one or more pair of genes.
Sequencing	Determining the order of nucleotides in a DNA or RNA molecule or the order of amino acids in a protein is sequencing.
Endangered species	An endangered species is a population of organisms (frequently but not always a taxonomic species) which is either (a) so few in number or (b) threatened by changing environmental or predation parameters that it is at risk of becoming extinct.
DNA sequencing	The process of determining the chemical composition of a DNA molecule is called DNA sequencing.
Trait	In biology, a trait or character is a genetically inherited feature of an organism: Eye color is a character; brown eyes and blue eyes is a trait.
Distribution	Distribution in pharmacology is a branch of pharmacokinetics describing reversible transfer of drug from one location to another within the body.
Egg	An egg is the zygote, resulting from fertilization of the ovum. It nourishes and protects the embryo.
Reproduction	Biological reproduction is the biological process by which new individual organisms are produced. Reproduction is a fundamental feature of all known life; each individual organism exists as the result of reproduction by an antecedent.
Somatic	The term somatic refers to the body. It also refers to the part of the nervous system that controls voluntary movement and senzation and judges relative effort and weight, called proprioception.
Carapace	Carapace refers to the shield-like structure that covers the anterior portion of some crustaceans. The shell of sea turtles.
Organ	Organ refers to a structure consisting of several tissues adapted as a group to perform specific functions.
Growth ring	A growth ring can be seen in a horizontal cross section cut through the trunk of a tree. Visible rings result from the change in growth speed through the seasons of the year, thus one ring usually marks the passage of one year in the life of the tree.
Mortality rate	Mortality rate is the number of deaths (from a disease or in general) per 1000 people and typically reported on an annual basis.
Growth rate	A measure of the change in population size per individual per unit of time is called growth rate.

Pulse	The rhythmic stretching of the arteries caused by the pressure of blood forced through the arteries by contractions of the ventricles during systole is a pulse.
Plankton	Plankton are drifting organisms that inhabit the water column of oceans, seas, and bodies of fresh water. Plankton abundance and distribution are strongly dependent on factors such as ambient nutrients concentrations, the physical state of the water column, and the abundance of other plankton.
Otolith	An otolith is a structure in the saccule or utricle of the inner ear. It is comprized of a combination of a gelatinous matrix and calcium carbonate crystals.
Habitat	Habitat refers to a place where an organism lives; an environmental situation in which an organism lives.
Threshold	Electrical potential level at which an action potential or nerve impulse is produced is called threshold.
Coral reef	A coral reef is a type of biotic reef that develops in tropical waters by the growth of coralline algae, hermatypic corals, and other marine organisms.
Gill	An extension of the body surface of an aquatic animal, specialized for gas exchange and/or suspension feeding is called a gill.
Nekton	Animals, such as fish, that can swim against currents of water are referred to as nekton.
Crustacea	The animal phylum that includes mostly aquatic animals such as crabs, lobsters, crayfish, and shrimp is referred to as crustacea.
Crab	The term crab is often applied to several different groups of short (nose to tail) decapod crustaceans with thick exoskeletons, but only members of the Brachyura are true.
Cone	Cone refers to a reproductive structure of gymnosperms that produces pollen in males or eggs in females.
Maximum sustainable yield	In fisheries science, maximum sustainable yield is the largest long-term average yield/catch that can be taken from a stock of fish without depressing the species' ability to reproduce.
Trophic structure	The different feeding relationships in an ecosystem, which determine the route of energy flow and the pattern of chemical cycling is called trophic structure.
Water column	The vertical column of seawater that extends from the surface to the bottom is called water column.
Overfishing	Overfishing is a situation where one or more fish stocks are reduced below predefined levels of acceptance by fishing activities. Biological overfishing occurs when fishing mortality has reached a level where the stock biomass has negative marginal growth .
Predator	A predator is an animal or other organism that hunts and kills other organisms for food in an act called predation.
Biomass	Biomass is organic non-fossil material, collectively. In other words, biomass comprises the mass of all biological organisms, dead or alive, excluding biological mass that has been transformed by geological processes into substances such as coal or petroleum.
Invertebrate	Invertebrate is a term coined by Jean-Baptiste Lamarck to describe any animal without a spinal column. It therefore includes all animals except vertebrates (fish, reptiles, amphibians, birds and mammals).
Pollution	Any environmental change that adversely affects the lives and health of living things is referred to as pollution.

Climate	Weather condition of an area including especially prevailing temperature and average daily/yearly rainfall over a long period of time is called climate.
Salinity	Salinity refers to the total amount of salts dissolved in seawater. It is generally expressed in parts per thousand.
Surface water	Includes all bodies of water, lakes, rivers, ponds, and so on that are on the surface of the earth, in contrast to groundwater, which lies below the surface is surface water.
Generation time	Generation time refers to the time between the birth of a parent and the birth of its offspring. The time interval between successive cell divisions.
Continental shelf	The shallow part of the seafloor immediately adjacent to a continent is called the continental shelf.
Sonar	A technique or equipment used to locate objects underwater by the detection of echoes is referred to as sonar.
Horizon	Horizon refers to levels within a soil profile that differ structurally and chemically. Generally divided into A, B, C, E, and 0 horizons.
Sperm	Sperm refers to the male sex cell with three distinct parts at maturity: head, middle piece, and tail.
Baleen whales	The distinguishing feature between this suborder and the toothed whales is that baleen whales have baleen plates instead of teeth in the upper jaw, which enables them to filter food from water.
Baleen	Baleen is the substance made of keratin that makes up the filtering plates that hang from the upper jaws of baleen whales.
Gestation	Gestation refers to pregnancy; the state of carrying developing young within the female reproductive tract.
Spout	Spout refers to the water vapor and seawater that is observed when whales surface and exhale.
Biology	Biology is the branch of science dealing with the study of life. It is concerned with the characteristics, classification, and behaviors of organisms, how species come into existence, and the interactions they have with each other and with the environment.
Dissolved oxygen	Oxygen gas molecules dissolved in water. Fish and other aquatic organisms are dependent on dissolved oxygen for respiration. Therefore concentration of dissolved oxygen is a measure of water quality.
Deforestation	The excessive cutting of forests, primarily rain forests in the Tropics, to clear space for agriculture is called deforestation.
Clear-cutting	Clear-cutting refers to the practice of cutting all trees in an area, regardless of species, size, quality, or age.
Erosion	Erosion is the displacement of solids (soil, mud, rock, and other particles) by the agents of wind, water, ice, movement in response to gravity, or living organisms.
Soil erosion	Soil erosion refers to the loss of soil caused by particles being carried away by wind and/or water.
Sediment	Sediment is any particulate matter that can be transported by fluid flow and which eventually is deposited as a layer of solid particles on the bed or bottom of a body of water or other liquid.
Salt marsh	Salt marsh refers to a grassy area that extends along the shores of estuaries and sheltered coasts in temperate and subpolar regions.

Bycatch	Bycatch refers to organisms accidentally caught and killed during a commercial fishing operation.
Catastrophe	An unpredictable event that has a strong effect on a population is referred to as a catastrophe.
Snake	A snake also known as ophidians, are cold blooded legless reptiles closely related to lizards, which share the order Squamata.
Upstream	Upstream refers to movement opposite the direction RNA follows when moving along a gene.
Environmental movement	The environmental movement is a diverse global social and political movement, which advocates for the protection, sustainable management and restoration of the natural environment in an effort to satisfy human needs, including spiritual and social needs, as well as for its own sake.
Echinoderm	Echinoderm is a phylum of marine animals found in the ocean at all depths. This phylum dates back to the lower Cambrian period and represents about 7000 living species and 13000 extinct ones. 6 classes made it to the Modern Era.
Epifauna	Epifauna are organisms that live upon the surface of sediments or soils.
Bryozoan	Colonial animals that build stony skeletons of calcium carbonate and superficially resemble coral is referred to as bryozoan.
Ecosystem	In general terms an ecosystem can be thought of as an assemblage of organisms (plant, animal and other living organisms living together with their environment, functioning as a loose unit. That is, a dynamic and complex whole, interacting as an "ecological unit".
Mammal	Homeothermic vertebrate characterized especially by the presence of hair and mammary glands is a mammal.
Metapopulation	A metapopulation consists of a group of spatially separated populations of the same species which interact at some level.
Source population	A population that can provide members to other populations of the species because it lives in a favorable area, and the birthrate is most likely higher than the death rate is called a source population.
Dispersal	Biological dispersal refers to those processes by which a species maintains or expands the distribution of a population. Dispersal implies movement—movement away from an existing population or away from the parent organisms.
Sink population	Population found in an unfavorable area where at best the birthrate equals the death rate is a sink population.
Thrush	Thrush is an infection of the oral mucous membrane by the fungus Candida albicans; also known as oral candidiasis.
Coral	Group of cnidarians having a calcium carbonate skeleton that participate in the formation of reefs is coral.
Substratum	Substratum is the underlying layer, or base to which a lichen is fixed.
Kelp	Kelp are large seaweeds, belonging to the brown algae and classified in the order Laminariales. Despite their appearance they are not grouped with the normal aquatic or land plants (kingdom Plantae), but instead are included in either kingdom Protista or Chromista.
Bleaching	Bleaching is a phenomenon where severely stressed corals turn white, frequently due to high water temperatures. This color change is mainly due to the corals expelling their zooxanthellae.

Starfish	Starfish are marine invertebrates belonging to phylum Echinodermata, class Asteroidea. They exhibit a superficially radial symmetry, typically with five or more "arms" which radiate from an indistinct disk. They do not have movable skeletons, but instead possess a hydraulic water vascular system.
Parasite	A parasite is an organism that spends a significant portion of its life in or on the living tissue of a host organism and which causes harm to the host without immediately killing it. They also commonly show highly specialized adaptations allowing them to exploit host resources.
Food chain	Food chain refers to a sequence of food transfers from one species to another within an ecosystem.
Polyculture	Polyculture refers to the growing of two or more species together. Contrasts with monoculture.
Monoculture	In agriculture, monoculture describes the practice of relying on a very small number of genetic variants, or cultivars of a food crop for commercial agricultural. Modern agriculture relies on standardization on a single cultivar so that the technology for tilling, planting, pest control, and harvesting, can be used over large geographical areas to obtain an economy of scale.
Growth form	One of several categories of the physical structure of plants, such as tree, herbaceous perennial, or liana is called growth form.
Mangrove	Mangrove are woody trees or shrubs that grow in coastal habitats or mangal (Hogarth, 1999), for which the term mangrove swamp also would apply.
Phytoplankton	Phytoplankton refers to the autotrophic component of the plankton that drifts in the water column. Phytoplankton obtain energy through a process called photosynthesis, and so must live in the well-lit surface layer (termed the euphotic zone) of an ocean, sea, or lake.
Mollusk	The mollusk is the large and diverse phylum Mollusca, which includes a variety of familiar creatures well-known for their decorative shells or as seafood. These range from tiny snails, clams, and abalone to the octopus, cuttlefish and squid (which are considered the most intelligent invertebrates).
Bivalve	Bivalve refers to a member of a group of mollusks that includes clams, mussels, scallops, and oysters. They typically have two-part shells, with both parts being more or less symmetrical.
Suspension feeder	An animal that extracts food particles suspended in the surrounding water for food is referred to as a suspension feeder.
Mussel	A mussel is a bivalve mollusk that can be found in lakes, rivers, creeks, intertidal areas, and throughout the ocean. The saltwater mussels (family Mytilidae) and freshwater mussels (family Unionidae) are not thought to be closely related and are grouped in different subclasses, despite considerable similarities in appearance.
Mudflat	Mudflat is a coastal wetland that forms when mud is deposited by the tides or rivers, they are found in sheltered areas such as bays, bayous, lagoons, and estuaries.
Metamorphosis	Metamorphosis is a process in biology by which an individual physically develops after birth or hatching, and involves significant change in form as well as growth and differentiation.
Snail	The name snail applies to most members of the molluscan class Gastropoda that have coiled shells. Other gastropods, which lack a conspicuous shell, are commonly called slugs, and are scattered throughout groups that primarily include snails.
Amoeboid	Amoeboid is a cell that moves or feeds by means of temporary projections, called pseudopods. They have appeared in a number of different groups. Some cells in multicellular animals may

	be amoeboid, for instance our white blood cells, which consume pathogens
Algal bloom	An algal bloom is a relatively rapid increase in the population of phytoplankton algae in an aquatic system. Typically only one or a few species are involved and the bloom is recognized by discoloration of the water resulting from the high density of pigmented cells.
Red tide	Red tide is estuarine or marine algal bloom. Red tide is caused by species of dinoflagellates, often present in sufficient numbers (thousands or millions of cells per milliliter) to turn the water red or brown.
Dinoflagellate	A unicellular photosynthetic alga with two flagella situated in perpendicular grooves in cellulose plates covering the cell is a dinoflagellate.
Toxin	Toxin refers to a microbial product or component that can injure another cell or organism at low concentrations. Often the term refers to a poisonous protein, but toxins may be lipids and other substances.
Paralytic shellfish poisoning	Dinoflagellates produce a powerful neurotoxin called saxitoxin. Shellfish accumulate saxitoxin and are poisonous when consumed by animals and humans. Saxitoxin paralyzes the striated respiratory muscles by inhibiting sodium transport. Paralytic shellfish poisoning is characterized by numbness of the mouth, lips, face, and extremities.
Genus	In biology, a genus is a taxonomic grouping. That is, in the classification of living organisms, a genus is considered to be distinct from other such genera. A genus has one or more species: if it has more than one species these are likely to be morphologically more similar than species belonging to different genera.
Spine	A spine is a rigid, pointed surface protuberance or needle-like structure on an animal, shell, or plant, presumably serving as a defense against attack by predators.
Tide	The periodic, rhythmic rise and fall of the sea surface caused by changes in gravitational forces external to the Earth is referred to as the tide.
Nerve cell	A cell specialized to originate or transmit nerve impulses is referred to as nerve cell.
Neurotoxin	Neurotoxin refers to a chemical that damages the nervous system of an animal.
Sodium	Sodium is the chemical element in the periodic table that has the symbol Na (Natrium in Latin) and atomic number 11. Sodium is a soft, waxy, silvery reactive metal belonging to the alkali metals that is abundant in natural compounds (especially halite). It is highly reactive.
Tissue	Group of similar cells which perform a common function is called tissue.
Diatom	Diatom refers to a unicellular photosynthetic alga with a unique, glassy cell wall containing silica.
Acid	An acid is a water-soluble, sour-tasting chemical compound that when dissolved in water, gives a solution with a pH of less than 7.
Lesion	A lesion is a non-specific term referring to abnormal tissue in the body. It can be caused by any disease process including trauma (physical, chemical, electrical), infection, neoplasm, metabolic and autoimmune.
Skin	Skin is an organ of the integumentary system composed of a layer of tissues that protect underlying muscles and organs.
Dementia	Dementia is progressive decline in cognitive function due to damage or disease in the brain beyond what might be expected from normal aging.
Cyst	A cyst is a closed sac having a distinct membrane and developing abnormally in a cavity or structure of the body. They may occur as a result of a developmental error in the embryo

	during pregnancy or they may be caused by infections.
Hot spots	Hot spots refers to sites within a gene that mutate more frequently than others, either spontaneously or after treatment with a particular mutagen.
Upwelling	Upward movement of deep, nutrient-rich water along coasts is referred to as upwelling.
Seaweed	Phycologists consider seaweed to refer any of a large number of marine benthic algae that are multicellular, macrothallic (large-bodied), and thus differentiated from most algae that tend to be microscopic in size.
Spore	Spore refers to a differentiated, specialized form that can be used for dissemination, for survival of adverse conditions because of its heat and dessication resistance, and/or for reproduction. They are usually unicellular and may develop into vegetative organisms or gametes. A spore may be produced asexually or sexually and are of many types.
Estuary	An estuary is a semi-enclosed coastal body of water which has a free connection with the open sea and within which sea water mixes with fresh water. The key feature of an estuary is that it is a mixing place for sea water and a stream or river to supply fresh water.
Thallus	Thallus is an undifferentiated vegetative tissue of some non-mobile organisms, which were previously known as the thallophytes.
Gamete	A gamete is a specialized germ cell that unites with another gamete during fertilization in organisms that reproduce sexually. They are haploid cells; that is, they contain one complete set of chromosomes. When they unite they form a zygote—a cell having two complete sets of chromosomes and therefore diploid.
Stipe	A stipe is a supportive structure that may be stemlike, as in seaweed, or a true leaf stem, as in ferns.
Gametophyte stage	A life cycle stage in plants in which a haploid sex cell is produced by mitosis is called gametophyte stage.
Sporophyte	A sporophyte is the diploid structure or phase of life of a sexually reproducing plant. Each living cell of the sporophyte contains two complete sets of chromosomes. The sporophyte is the dominant life form in ferns, gymnosperms, and angiosperms.
Crop	An organ, found in both earthworms and birds, in which ingested food is temporarily stored before being passed to the gizzard, where it is pulverized is the crop.
Vertebrate	Vertebrate is a subphylum of chordates, specifically, those with backbones or spinal columns. They started to evolve about 530 million years ago during the Cambrian explosion, which is part of the Cambrian period.
Microhabitat	A microhabitat is the immediate surroundings and other physical factors of an individual plant or animal within its habitat.
Fry	A fry is a recently-hatched fish, but one which has fully absorbed its yolk sac and can now hunt and consume live food.
Genetic diversity	Genetic diversity is a characteristic of ecosystems and gene pools that describes an attribute which is commonly held to be advantageous for survival -- that there are many different versions of otherwise similar organisms.
Selective breeding	Selective breeding refers to the mating of those members of a strain of animals or plants that manifest a particular characteristic, which may or may not be done deliberately, to affect the genetic makeup of future generations of that strain.
Gene	Gene refers to a discrete unit of hereditary information consisting of a specific nucleotide sequence in DNA . Most of the genes of a eukaryote are located in its chromosomal DNA; a few

	are carried by the DNA of mitochondria and chloroplasts.
Algae	The algae consist of several different groups of living organisms that capture light energy through photosynthesis, converting inorganic substances into simple sugars with the captured energy.
Aquaculture	Aquaculture is the cultivation of the natural produce of water (such as fish or shellfish, algae and other aquatic plants). Mariculture is specifically marine aquaculture, and thus is a subset of aquaculture.
Ecology	Ecology is the scientific study of the distribution and abundance of living organisms and how these properties are affected by interactions between the organisms and their environment.
Intertidal zone	The part of a beach that lies between average high tide and average low tide is called the intertidal zone.
Barrier reef	A type of coral reef that develops at some distance from the coast creating a lagoon between itself and the coast is called a barrier reef.
Oligonucleotide	Oligonucleotide refers to a short single-stranded DNA molecule, typically with twenty or fewer base pairs. Oligonucleotides are often used as probes for detecting complementary DNA or RNA because they bind readily to their complements.
Antibody	An antibody is a protein used by the immune system to identify and neutralize foreign objects like bacteria and viruses. Each antibody recognizes a specific antigen unique to its target.
Biotransform-tion	Alteration of the structure of a compound by a living organism or enzyme is biotransformation.
Hypoxia	Hypoxia is a lack of oxygen to the brain and other vital organs. This is treated by providing the patient with oxygen, either through a bag-valve-mask device, or by inserting an endotracheal tube (intubation).
Crustacean	Member of a group of marine arthropods that contains among others shrimps, crabs, crayfish, and lobsters is called crustacean.
Epidemic	An epidemic is a disease that appears as new cases in a given human population, during a given period, at a rate that substantially exceeds what is "expected", based on recent experience.
Amino acid	An amino acid is any molecule that contains both amino and carboxylic acid functional groups. They are the basic structural building units of proteins. They form short polymer chains called peptides or polypeptides which in turn form structures called proteins.

Pollution	Any environmental change that adversely affects the lives and health of living things is referred to as pollution.
Habitat	Habitat refers to a place where an organism lives; an environmental situation in which an organism lives.
Pollutant	Pollutant refers to any natural or artificial substance that enters the ecosystem in such quantities that it does harm to the ecosystem; any introduced substance that makes a resource unfit for a specific purpose.
Biotic	Biotic refers to life or living organisms.
Climate	Weather condition of an area including especially prevailing temperature and average daily/yearly rainfall over a long period of time is called climate.
Biota	Biota is the plant and animal life of a region or area or geological period. The biota of the Earth is known as the biosphere.
Point source	A source of a pollutant such as discharges from factory drains or outlets from sewage treatment plants is a point source.
Runoff	Water entering rivers, lakes, reservoirs, or the ocean from land surfaces is runoff.
Species	Group of similarly constructed organisms capable of interbreeding and producing fertile offspring is a species.
Population	Group of organisms of the same species occupying a certain area and sharing a common gene pool is referred to as population.
Relative abundance	Differences in the abundance of species within a community is relative abundance.
Substratum	Substratum is the underlying layer, or base to which a lichen is fixed.
Species richness	Species richness is perhaps the simplest measure of biodiversity. The larger the number, the more species in an area. It is referred to in equations as S. There is a strong inverse correlation in many groups between species richness and latitude - the further from the equator you go, the fewer species you will find.
Hydrocarbon	A chemical compound composed only of the elements carbon and hydrogen is called hydrocarbon.
Predator	A predator is an animal or other organism that hunts and kills other organisms for food in an act called predation.
Food web	A network of interconnecting food chains is referred to as a food web.
Sediment	Sediment is any particulate matter that can be transported by fluid flow and which eventually is deposited as a layer of solid particles on the bed or bottom of a body of water or other liquid.
Gradient	Gradient refers to a difference in concentration, pressure, or electrical charge between two regions.
Fjord	A fjord is a narrow inlet of the sea between cliffs or steep slopes, which results from marine inundation of a glaciated valley. Typical characteristics of a fjord include: a narrow inlet, a bottom glacially eroded significantly below sea level, steep-sided walls which continue to descend below the sea surface, greater depths in the upper and middle reaches than on the seaward side, and communication with the open sea.
Reproduction	Biological reproduction is the biological process by which new individual organisms are produced. Reproduction is a fundamental feature of all known life; each individual organism exists as the result of reproduction by an antecedent.

Crab	The term crab is often applied to several different groups of short (nose to tail) decapod crustaceans with thick exoskeletons, but only members of the Brachyura are true.
Mortality rate	Mortality rate is the number of deaths (from a disease or in general) per 1000 people and typically reported on an annual basis.
Vas deferens	Sperm are transferred from the vas deferens, muscular tubes, into the urethra, collecting fluids from the male accessory sex glands en route.
Snail	The name snail applies to most members of the molluscan class Gastropoda that have coiled shells. Other gastropods, which lack a conspicuous shell, are commonly called slugs, and are scattered throughout groups that primarily include snails.
Oligochaete	The Oligochaete or "few-bristled" worms are well-segmented Annelids, most with a spacious coelom that is used as a hydroskeleton. Their setae (chaetae) or "bristles" are generally few in number and they lack the parapodia of the polychaeta.
Trait	In biology, a trait or character is a genetically inherited feature of an organism: Eye color is a character; brown eyes and blue eyes is a trait.
Toxin	Toxin refers to a microbial product or component that can injure another cell or organism at low concentrations. Often the term refers to a poisonous protein, but toxins may be lipids and other substances.
Catastrophe	An unpredictable event that has a strong effect on a population is referred to as a catastrophe.
Pulse	The rhythmic stretching of the arteries caused by the pressure of blood forced through the arteries by contractions of the ventricles during systole is a pulse.
Hot spots	Hot spots refers to sites within a gene that mutate more frequently than others, either spontaneously or after treatment with a particular mutagen.
Erosion	Erosion is the displacement of solids (soil, mud, rock, and other particles) by the agents of wind, water, ice, movement in response to gravity, or living organisms.
Trophic level	In ecology, the trophic level is the position that an organism occupies in a food chain - what it eats, and what eats it.
Food chain	Food chain refers to a sequence of food transfers from one species to another within an ecosystem.
Carnivore	An animal that eats a diet consisting solely of meat is referred to as a carnivore.
Herbivore	A herbivore is an animal that is adapted to eat primarily plant matter
Tissue	Group of similar cells which perform a common function is called tissue.
Fungicide	Fungicide is a chemical compound used to prevent the spread of fungi in gardens and crops, which can cause serious damage to the plants. They are also used to fight fungal infections.
Copper	Copper is a chemical element in the periodic table that has the symbol Cu (L.: Cuprum) and atomic number 29. It is a ductile metal with excellent electrical conductivity, and finds extensive use as a building material, as an electrical conductor, and as a component of various alloys.
Denature	Denature is a structural change in biomolecules such as nucleic acids and proteins, such that they are no longer in their native state, and their shape which allows for optimal activity.
Protein	A protein is a complex, high-molecular-weight organic compound that consists of amino acids joined by peptide bonds. They are essential to the structure and function of all living cells and viruses. Many are enzymes or subunits of enzymes.

Zinc	Zinc is a chemical element in the periodic table that has the symbol Zn and atomic number 30.
Pigment	Pigment is any material resulting in color in plant or animal cells which is the result of selective absorption.
Blood	Blood is a circulating tissue composed of fluid plasma and cells. The main function of blood is to supply nutrients (oxygen, glucose) and constitutional elements to tissues and to remove waste products.
Organelle	In cell biology, an organelle is one of several structures with specialized functions, suspended in the cytoplasm of a eukaryotic cell.
Chemical reaction	Chemical reaction refers to a process leading to chemical changes in matter; involves the making and/or breaking of chemical bonds.
Cell	The cell is the structural and functional unit of all living organisms, and is sometimes called the "building block of life."
Placenta	The placenta is an organ present only in female placental mammals during gestation. It is composed of two parts, one genetically and biologically part of the fetus, the other part of the mother. It is implanted in the wall of the uterus, where it receives nutrients and oxygen from the mother's blood and passes out waste.
Kidney	The kidney is a bean-shaped excretory organ in vertebrates. Part of the urinary system, the kidneys filter wastes (especially urea) from the blood and excrete them, along with water, as urine.
Sulfide	One of the minerals that is abundant in the hot water that seeps through hydrothermal vents is sulfide.
Fossil fuel	Fossil fuel refers to an energy deposit formed from the remains of extinct organisms.
Nervous system	The nervous system of an animal coordinates the activity of the muscles, monitors the organs, constructs and processes input from the senses, and initiates actions.
Ecosystem	In general terms an ecosystem can be thought of as an assemblage of organisms (plant, animal and other living organisms living together with their environment, functioning as a loose unit. That is, a dynamic and complex whole, interacting as an "ecological unit".
Mussel	A mussel is a bivalve mollusk that can be found in lakes, rivers, creeks, intertidal areas, and throughout the ocean. The saltwater mussels (family Mytilidae) and freshwater mussels (family Unionidae) are not thought to be closely related and are grouped in different subclasses, despite considerable similarities in appearance.
Insect	An arthropod that usually has three body segments , three pairs of legs, and one or two pairs of wings is called an insect. They are the largest and (on land) most widely-distributed taxon within the phylum Arthropoda. They comprise the most diverse group of animals on the earth, with around 925,000 species described
Crop	An organ, found in both earthworms and birds, in which ingested food is temporarily stored before being passed to the gizzard, where it is pulverized is the crop.
Evolution	In biology, evolution is the process by which novel traits arise in populations and are passed on from generation to generation. Its action over large stretches of time explains the origin of new species and ultimately the vast diversity of the biological world.
Herbicide	Herbicide refers to a chemical substance that kills plants.
Solubility	A solution at equilibrium that cannot hold any more solute is said to be saturated. The equilibrium of a solution is mainly dependent on temperature. The maximum equilibrium amount of solute which can normally dissolve per amount of solvent is the solubility of that solute

	in that solvent.
Chlorinated hydrocarbon	Chlorinated hydrocarbon is a broad class of organic chemicals used mainly as solvents but also with many other uses. A chlorinated hydrocarbon is derived from a hydrocarbon molecule where one or more of the hydrogen atoms has been replaced by a chlorine atom.
Salt marsh	Salt marsh refers to a grassy area that extends along the shores of estuaries and sheltered coasts in temperate and subpolar regions.
Malaria	Malaria refers to potentially fatal human disease caused by the protozoan parasite Plasmodium, which is transmitted by the bite of an infected mosquito.
Residue	A residue refers to a portion of a larger molecule, a specific monomer of a polysaccharide, protein or nucleic acid.
Egg	An egg is the zygote, resulting from fertilization of the ovum. It nourishes and protects the embryo.
Biology	Biology is the branch of science dealing with the study of life. It is concerned with the characteristics, classification, and behaviors of organisms, how species come into existence, and the interactions they have with each other and with the environment.
Overfishing	Overfishing is a situation where one or more fish stocks are reduced below predefined levels of acceptance by fishing activities. Biological overfishing occurs when fishing mortality has reached a level where the stock biomass has negative marginal growth .
Gill	An extension of the body surface of an aquatic animal, specialized for gas exchange and/or suspension feeding is called a gill.
Diatom	Diatom refers to a unicellular photosynthetic alga with a unique, glassy cell wall containing silica.
Tide	The periodic, rhythmic rise and fall of the sea surface caused by changes in gravitational forces external to the Earth is referred to as the tide.
Crustacean	Member of a group of marine arthropods that contains among others shrimps, crabs, crayfish, and lobsters is called crustacean.
Watershed	The total land area that drains directly or indirectly into a particular stream or river. The watershed is generally named for the stream or river into which it drains.
Invertebrate	Invertebrate is a term coined by Jean-Baptiste Lamarck to describe any animal without a spinal column. It therefore includes all animals except vertebrates (fish, reptiles, amphibians, birds and mammals).
Dioxin	The group of toxic chlorinated hydrocarbon pollutants is called dioxin.
Polychlorinated biphenyl	Polychlorinated biphenyl is a class of organic compounds with 1 to 10 chlorine atoms attached to biphenyl and a general structure of $C_{12}H_{10-x}Cl_x$.
Mammal	Homeothermic vertebrate characterized especially by the presence of hair and mammary glands is a mammal.
Carcinoma	Cancer that originates in the coverings of the body, such as the skin or the lining of the intestinal tract is a carcinoma.
Bacteria	The domain that contains procaryotic cells with primarily diacyl glycerol diesters in their membranes and with bacterial rRNA. Bacteria also is a general term for organisms that are composed of procaryotic cells and are not multicellular.
Benthos	Benthos are the organisms and habitats of the sea floor; in freshwater biology they are the organisms and habitats of the bottoms of lakes, rivers, and creeks.

Amphipods	Amphipods include over 7000 described species of small, shrimp-like crustaceans. Most amphipods are marine, although a few live in freshwater or are terrestrial. One species common to mid-Atlantic and estuarine waters is Jassa falcata.
Aromatic hydrocarbon	An aromatic hydrocarbon, or arene is a hydrocarbon, the molecular structure of which incorporates one or more planar sets of six carbon atoms that are connected by delocalized electrons numbering the same as if they consisted of alternating single and double covalent bonds.
Benthic community	The benthic community is an assemblage of interacting organisms, including those forming structural elements, found at or near the bottom of a body of water.
Anadromous	Anadromous refers to marine fishes that migrate to fresh water to breed.
Mangrove	Mangrove are woody trees or shrubs that grow in coastal habitats or mangal (Hogarth, 1999), for which the term mangrove swamp also would apply.
Rhizome	A rhizome is a usually-underground, horizontal stem of a plant that often sends out roots and shoots from its nodes.
Coral	Group of cnidarians having a calcium carbonate skeleton that participate in the formation of reefs is coral.
Coral reef	A coral reef is a type of biotic reef that develops in tropical waters by the growth of coralline algae, hermatypic corals, and other marine organisms.
Alkane	An alkane in organic chemistry is a saturated hydrocarbon without cycles, that is, an acyclic hydrocarbon in which the molecule has the maximum possible number of hydrogen atoms and so has no double bonds. The general formula is C_nH_{2n+2}.
Hydrogen atom	A hydrogen atom is an atom of the element hydrogen. It is composed of a single negatively-charged electron, attending a positively-charged proton which is the nucleus of the hydrogen atom. The electron is bound to the proton by the Coulomb force.
Double bond	Double bond refers to a type of covalent bond in which two atoms share two pairs of electrons; symbolized by a pair of lines between the bonded atoms. An example is in ethylene (between the carbon atoms). It usually consists of one sigma bond and one pi bond.
Benzene	Benzene is an organic chemical compound that is a colorless and flammable liquid with a pleasant, sweet smell. Benzene is a known carcinogen. It is a minor, or additive, component of gasoline. It is an important industrial solvent and precursor in the production of drugs, plastics, gasoline, synthetic rubber, and dyes.
Aromatic	In chemistry, an aromatic molecule is one in which a conjugated ring of unsaturated bonds, lone pairs, or empty orbitals exhibit a stabilization stronger than would be expected by the stabilization of conjugation alone.
Cancer	Cancer is a class of diseases or disorders characterized by uncontrolled division of cells and the ability of these cells to invade other tissues, either by direct growth into adjacent tissue through invasion or by implantation into distant sites by metastasis.
Alkene	An alkene in organic chemistry is an unsaturated hydrocarbon containing at least one carbon to carbon double bond. The simplest alkene, with only one double bond, forms a homologous series, the alkene with general formula C_nH_{2n}.
Carbon	Carbon is a chemical element in the periodic table that has the symbol C and atomic number 6. An abundant nonmetallic, tetravalent element, carbon has several allotropic forms.
Atom	An atom is the smallest possible particle of a chemical element that retains its chemical properties.

Cell membrane	A component of every biological cell, the selectively permeable cell membrane is a thin and structured bilayer of phospholipid and protein molecules that envelopes the cell. It separates a cell's interior from its surroundings and controls what moves in and out.
Hydrophobic	Hydrophobic refers to being electrically neutral and nonpolar, and thus prefering other neutral and nonpolar solvents or molecular environments. Hydrophobic is often used interchangeably with "oily" or "lipophilic."
Dispersal	Biological dispersal refers to those processes by which a species maintains or expands the distribution of a population. Dispersal implies movement—movement away from an existing population or away from the parent organisms.
Hydrophilic	A hydrophilic molecule or portion of a molecule is one that is typically charge-polarized and capable of hydrogen bonding, enabling it to dissolve more readily in water than in oil or other hydrophobic solvents.
Surfactant	A substance that decreases the surface tension of a liquid, allowing easier spreading, and the interfacial tension between two liquids is a surfactant.
Molecule	A molecule is the smallest particle of a pure chemical substance that still retains its chemical composition and properties.
Mass spectrometry	Mass spectrometry is an analytical technique used to measure the mass-to-charge ratio (m/q) of ions. It is most generally used to find the composition of a physical sample by generating a mass spectrum representing the masses of sample components.
Chromatography	Biochemical technique in which a mixture of substances is separated by charge, size, or some other property by allowing it to partition between a mobile phase and a stationary phase is called chromatography.
Molecular weight	The molecular mass of a substance, called molecular weight and abbreviated as MW, is the mass of one molecule of that substance, relative to the unified atomic mass unit u (equal to 1/12 the mass of one atom of carbon-12).
Vertebrate	Vertebrate is a subphylum of chordates, specifically, those with backbones or spinal columns. They started to evolve about 530 million years ago during the Cambrian explosion, which is part of the Cambrian period.
Gonadotropin-releasing hormone	Gonadotropin-releasing hormone is a peptide hormone responsible for the release of FSH and LH from the anterior pituitary. It is synthesized and released by the hypothalamus. It is considered a neurohormone, a hormone produced in a specific neural cell and released at its neural terminal.
Ovarian follicle	Ovarian follicle is the roughly spherical cell aggregation in the ovary containing an ovum and from which the egg is released during ovulation.
Pituitary gland	The pituitary gland is an endocrine gland about the size of a pea that sits in the small, bony cavity (sella turcica) at the base of the brain. Its posterior lobe is connected to a part of the brain called the hypothalamus via the infundibulum (or stalk), giving rise to the tuberoinfundibular pathway.
Hypothalamus	Located below the thalamus, the hypothalamus links the nervous system to the endocrine system by synthesizing and secreting neurohormones often called releasing hormones because they function by stimulating the secretion of hormones from the anterior pituitary gland.
Estradiol	Estradiol is a sex hormone. Labelled the "female" hormone but also present in males it represents the major estrogen in humans. Critical for sexual functioning estradiol also supports bone growth.
Secretion	Secretion is the process of segregating, elaborating, and releasing chemicals from a cell, or

a secreted chemical substance or amount of substance.

Lipoprotein

A lipoprotein is a biochemical assembly that contains both proteins and lipids and may be structural or catalytic in function. They may be enzymes, proton pumps, ion pumps, or some combination of these functions.

Oviparous

In some animals, an egg is the zygote, resulting from fertilization of the ovum. It nourishes and protects the embryo. Oviparous animals are animals that lay eggs, with little or no other development within the mother.

Liver

The liver is an organ in vertebrates, including humans. It plays a major role in metabolism and has a number of functions in the body including drug detoxification, glycogen storage, and plasma protein synthesis. It also produces bile, which is important for digestion.

Ovary

In the flowering plants, an ovary is a part of the female reproductive organ of the flower or gynoecium.

Yolk

Dense nutrient material that is present in the egg of a bird or reptile is referred to as yolk.

Hormone

A hormone is a chemical messenger from one cell to another. All multicellular organisms produce hormones. The best known hormones are those produced by endocrine glands of vertebrate animals, but hormones are produced by nearly every organ system and tissue type in a human or animal body. Hormone molecules are secreted directly into the bloodstream, they move by circulation or diffusion to their target cells, which may be nearby cells in the same tissue or cells of a distant organ of the body.

Gene expression

Gene expression is the process by which a gene's information is converted into the structures and functions of a cell. Gene expression is a multi-step process that begins with transcription, post transcriptional modification and translation, followed by folding, post-translational modification and targeting.

Cytochrome

Cytochrome refers to colored, heme-containing protein that transfers electrons during cellular respiration and photosynthesis.

Eutrophication

Eutrophication is the enrichment of an ecosystem with chemical nutrients, typically compounds containing nitrogen, phosphorus, or both.

Detoxification

Detoxification in general is the removal of toxic substances from the body. It is one of the functions of the liver and kidneys, but can also be achieved artificially by techniques such as dialysis and (in a very limited number of cases) chelation therapy.

Gene

Gene refers to a discrete unit of hereditary information consisting of a specific nucleotide sequence in DNA . Most of the genes of a eukaryote are located in its chromosomal DNA; a few are carried by the DNA of mitochondria and chloroplasts.

Microorganism

A microorganism is an organism that is so small that it is microscopic (invisible to the naked eye). They are often illustrated using single-celled, or unicellular organisms; however, some unicellular protists are visible to the naked eye, and some multicellular species are microscopic.

Suspension feeder

An animal that extracts food particles suspended in the surrounding water for food is referred to as a suspension feeder.

Salmonella

Salmonella is a genus of rod-shaped Gram-negative enterobacteria that causes typhoid fever, paratyphoid and foodborne illness. It is motile in nature and produces hydrogen sulfide.

Bacterium

Most bacterium are microscopic and unicellular, with a relatively simple cell structure lacking a cell nucleus, and organelles such as mitochondria and chloroplasts. They are the most abundant of all organisms. They are ubiquitous in soil, water, and as symbionts of other

	organisms.
Hepatitis	Hepatitis is a gastroenterological disease, featuring inflammation of the liver. The clinical signs and prognosis, as well as the therapy, depend on the cause.
Pathogen	A pathogen or infectious agent is a biological agent that causes disease or illness to its host.The term is most often used for agents that disrupt the normal physiology of a multicellular animal or plant.
Virus	Obligate intracellular parasite of living cells consisting of an outer capsid and an inner core of nucleic acid is referred to as virus. The term virus usually refers to those particles that infect eukaryotes whilst the term bacteriophage or phage is used to describe those infecting prokaryotes.
Coliform	Coliform bacteria are defined as rod-shaped Gram-negative organisms which ferment lactose with the production of gas when incubated at 35 °C.
Nutrient source	A habitat that releases more nutrients than flow into it is referred to as a nutrient source.
Limiting factor	In biology, agricultural science, physiology, and ecology, a limiting factor is one that controls a process, such as organism growth or species population size or distribution. The availability of food, predation pressure, or availability of shelter are examples of factors that could be limiting for a species population in a specific area.
Phytoplankton	Phytoplankton refers to the autotrophic component of the plankton that drifts in the water column. Phytoplankton obtain energy through a process called photosynthesis, and so must live in the well-lit surface layer (termed the euphotic zone) of an ocean, sea, or lake.
Atmosphere	Earth's atmosphere is a layer of gases surrounding the planet Earth and retained by the Earth's gravity. It contains roughly 78% nitrogen and 21% oxygen, with trace amounts of other gases.
Nitrogen	A colorless and tasteless and mostly inert diatomic non-metal gas that is an essential constituent of proteins is nitrogen.
Acid rain	Acid rain is defined as any type of precipitation with a pH of below 5.0. Acid rain is caused by sulfur from volcanic venting or impurities in fossil fuels, and nitrogen from the air, combining with oxygen to form sulfur dioxide and nitrogen oxides.
Sulfur	Sulfur is the chemical element in the periodic table that has the symbol S and atomic number 16. It is an abundant, tasteless, odorless, multivalent non-metal. Sulfur, in its native form, is a yellow crystaline solid. In nature, it can be found as the pure element or as sulfide and sulfate minerals.
Nitrate	Nitrate refers to a salt of nitric acid; a compound containing the radical NO_3; biologically, the final form of nitrogen from the oxidation of organic nitrogen compounds.
Primary production	Primary production is the production of organic compounds from inorganic materials principally through the process of photosynthesis.
Hypoxia	Hypoxia is a lack of oxygen to the brain and other vital organs. This is treated by providing the patient with oxygen, either through a bag-valve-mask device, or by inserting an endotracheal tube (intubation).
Zooplankton	Nonphotosynthetic protists that feed on other plankton and are abundant in marine and freshwater environments are zooplankton.
Water column	The vertical column of seawater that extends from the surface to the bottom is called water column.
Dissolved oxygen	Oxygen gas molecules dissolved in water. Fish and other aquatic organisms are dependent on

	dissolved oxygen for respiration. Therefore concentration of dissolved oxygen is a measure of water quality.
Oxygen	Oxygen is a chemical element in the periodic table. It has the symbol O and atomic number 8. Oxygen is the second most common element on Earth, composing around 46% of the mass of Earth's crust and 28% of the mass of Earth as a whole, and is the third most common element in the universe.
Hydrogen sulfide	The gas that is produced in anoxic sediments is referred to as hydrogen sulfide.
Continental shelf	The shallow part of the seafloor immediately adjacent to a continent is called the continental shelf.
Decomposition	Decomposition refers to the reduction of the body of a formerly living organism into simpler forms of matter.
Secondary treatment	Secondary treatment refers to the biological degradation of dissolved organic matter in the process of sewage treatment; the organic material is either mineralized or changed to settleable solids.
Primary treatment	The first step of sewage treatment, in which physical settling and screening are used to remove particulate materials is referred to as primary treatment.
Organic compound	An organic compound is any member of a large class of chemical compounds whose molecules contain carbon, with the exception of carbides, carbonates, carbon oxides and gases containing carbon.
Phosphate	A phosphate is a polyatomic ion or radical consisting of one phosphorus atom and four oxygen. In the ionic form, it carries a -3 formal charge, and is denoted PO_4^{3-}.
Colloid	Colloid refers to a mixture that contains dispersed particles larger than molecules but small enough so that they do not settle out.
Ammonia	Ammonia is a compound of nitrogen and hydrogen with the formula NH_3. At standard temperature and pressure ammonia is a gas. It is toxic and corrosive to some materials, and has a characteristic pungent odor.
Tertiary treatment	Tertiary treatment refers to the removal from sewage of inorganic nutrients, heavy metals, viruses, etc., by chemical and biological means after microorganisms have degraded dissolved organic material during secondary sewage treatment.
Anaerobic	An anaerobic organism is any organism that does not require oxygen for growth.
Iron	Iron is essential to all organisms, except for a few bacteria. It is mostly stably incorporated in the inside of metalloproteins, because in exposed or in free form it causes production of free radicals that are generally toxic to cells.
Phosphorus	Phosphorus is the chemical element in the periodic table that has the symbol P and atomic number 15.
Denitrification	Conversion of nitrate or nitrite to nitrogen gas by bacteria in soil is denitrification.
Sludge	Sludge refers to a general term for the precipitated solid matter produced during water and sewage treatment; solid particles composed of organic matter and microorganisms that are involved in aerobic sewage treatment .
Digestion	Digestion refers to the mechanical and chemical breakdown of food into molecules small enough for the body to absorb; the second main stage of food processing, following ingestion.
Compost	Compost refers to a mixture of decomposed organic matter, particularly decomposed plant materials .

Thermal pollution	Thermal pollution is a tempurature change in water caused by human influence. It is most significant with respect to discharging waste heat from electric generating plants, especially nuclear power plants, into bodies of water.
Cooling tower	A tower designed to dissipate waste heat from a power plant into the atmosphere is a cooling tower.
Closed system	An ecosystem that can can exchange heat and work, but not matter, with its surroundings is called a closed system.
Estuary	An estuary is a semi-enclosed coastal body of water which has a free connection with the open sea and within which sea water mixes with fresh water. The key feature of an estuary is that it is a mixing place for sea water and a stream or river to supply fresh water.
Entrainment	Entrainment is the process whereby two connected oscillating systems, having similar periods, fall into synchrony.
Algae	The algae consist of several different groups of living organisms that capture light energy through photosynthesis, converting inorganic substances into simple sugars with the captured energy.
Cyanobacteria	Cyanobacteria are a phylum of bacteria that obtain their energy through photosynthesis. They are often referred to as blue-green algae, even though it is now known that they are not directly related to any of the other algal groups, which are all eukaryotes.
Biomass	Biomass is organic non-fossil material, collectively. In other words, biomass comprises the mass of all biological organisms, dead or alive, excluding biological mass that has been transformed by geological processes into substances such as coal or petroleum.
Global warming	A slow but steady rise in Earth's surface temperature, caused by increasing concentrations of greenhouse gases in the atmosphere is referred to as global warming.
Carbon dioxide	Carbon dioxide is an atmospheric gas comprized of one carbon and two oxygen atoms. A very widely known chemical compound, it is frequently called by its formula CO_2. In its solid state, it is commonly known as dry ice.
Deforestation	The excessive cutting of forests, primarily rain forests in the Tropics, to clear space for agriculture is called deforestation.
Tropical forest	A tropical forest is a tropical and subtropical biome. They are moist broadleaf forests found in a belt around the equator and in the humid subtropics, and are characterized by warm, humid climates with high year-round rainfall.
Photosynthesis	Photosynthesis is a biochemical process in which plants, algae, and some bacteria harness the energy of light to produce food. Ultimately, nearly all living things depend on energy produced from photosynthesis for their nourishment, making it vital to life on Earth.
Respiration	Respiration is the process by which an organism obtains energy by reacting oxygen with glucose to give water, carbon dioxide and ATP (energy). Respiration takes place on a cellular level in the mitochondria of the cells and provide the cells with energy.
Greenhouse effect	The warming of the atmosphere caused by COD CH4, and other gases that absorb infrared radiation and slow its escape from Earth's surface is called greenhouse effect.
Radiation	The emission of electromagnetic waves by all objects warmer than absolute zero is referred to as radiation.
Sunlight	Sunlight in the broad sense is the total spectrum of electromagnetic radiation given off by the Sun.
Chlorofluoro-	A chlorofluorocarbon is a haloalkane with both chlorine and fluorine. They were formerly used

arbon	widely in industry, for example as refrigerants, propellants, and cleaning solvents.
Ozone layer	The ozone layer is that part of the Earth's atmosphere which contains relatively high concentrations of ozone (O_3). "Relatively high" means a few parts per million - much higher than the concentrations in the lower atmosphere but still small compared to the main components of the atmosphere.
Greenhouse gas	Greenhouse gas refers to a gas, such as carbon dioxide or methane, that traps sunlight energy in a planet's atmosphere as heat; a gas that participates in the greenhouse effect.
Prediction	Step of the scientific method which follows the formulation of a hypothesis and assists in creating the experimental design is referred to as prediction.
Coral bleaching	Coral bleaching refers to the loss of color from a coral as it expels its zooxanthellae. It is usually a stress response.
Upwelling	Upward movement of deep, nutrient-rich water along coasts is referred to as upwelling.
Surface water	Includes all bodies of water, lakes, rivers, ponds, and so on that are on the surface of the earth, in contrast to groundwater, which lies below the surface is surface water.
Desert	A desert is a landscape form or region that receives little precipitation - less than 250 mm (10 in) per year. It is a biome characterized by organisms adapted to sparse rainfall and rapid evaporation.
Nitrogen-fixing bacteria	Microorganisms that restock nitrogenous minerals in the soil by converting nitrogen to ammonia are referred to as nitrogen-fixing bacteria.
Primary productivity	Primary productivity is the amount of light energy converted to chemical energy in a given amount of time. It usually corresponds to the amount of photosynthesis that occurred.
Carbon cycle	Continuous process by which carbon circulates in the air, water, and organisms of the biosphere is the carbon cycle.
Chlorophyll A	A green pigment in chloroplasts that participates directly in the light reactions is referred to as chlorophyll A. Due to its central role in the reaction center, is essential for photosynthesis.
Gyre	Gyre refers to a roughly circular pattern of ocean currents, formed because continents interrupt the currents' flow; rotates clockwise in the Northern Hemisphere and counterclockwise in the Southern Hemisphere.
Thermocline	The thermocline is a layer within a body of water where the temperature changes rapidly with depth. Because water is not perfectly transparent, almost all sunlight is absorbed in the surface layer, which heats up.
Secchi disk	The Secchi disk is a device used to measure water transparency in open waters of lakes, bays, and the ocean.
Ecological impact	The total effect of an environmental change, whether natural or human is referred to as ecological impact.
Carbonic anhydrase	Carbonic anhydrase is a family of zinc-containing enzymes that catalyze the rapid interconversion of carbon dioxide and water into carbonic acid, protons, and bicarbonate ions.
Gland	A gland is an organ in an animal's body that synthesizes a substance for release such as hormones, often into the bloodstream or into cavities inside the body or its outer surface.
Gastropod	A member of the largest group of mollusks, including snails and slugs is a gastropod.
Absorption	Absorption is a physical or chemical phenomenon or a process in which atoms, molecules, or

	ions enter some bulk phase - gas, liquid or solid material. In nutrition, amino acids are broken down through digestion, which begins in the stomach.
Ecology	Ecology is the scientific study of the distribution and abundance of living organisms and how these properties are affected by interactions between the organisms and their environment.
Archipelago	An archipelago is a landform which consists of a chain or cluster of islands. They usually occur in the open sea; less commonly, a large land mass may neighbor them, and are often volcanic, forming along mid-ocean ridges or hotspots, but there are many other processes involved in their construction, including erosion and deposition.
Environmental toxicology	The study of the distribution and effects of toxic compounds in the environment is referred to as environmental toxicology.
Aquatic ecosystem	An aquatic ecosystem is based in water, whether it is a pond, lake, river, underground water body, estuary or ocean. It comprises living aquatic organisms which constitute the biotic factors of the ecosystem, as well as their physical environment which collectively can be referred to as the abiotic factor.
Heavy metals	The heavy metals are a group of elements between copper and lead on the periodic table of the elements—having atomic weights between 63.546 and 200.590 and specific gravities greater than
Adaptation	A biological adaptation is an anatomical structure, physiological process or behavioral trait of an organism that has evolved over a period of time by the process of natural selection such that it increases the expected long-term reproductive success of the organism.
Metabolism	Metabolism is the biochemical modification of chemical compounds in living organisms and cells. This includes the biosynthesis of complex organic molecules (anabolism) and their breakdown (catabolism).
Anatomy	Anatomy is the branch of biology that deals with the structure and organization of living things. It can be divided into animal anatomy (zootomy) and plant anatomy (phytonomy).
Ultraviolet radiation	Ultraviolet radiation refers to radiation similar to light but with wavelengths slightly shorter than violet light and with more energy. The greater energy causes it to severely burn and otherwise damage biological tissues.
Ozone depletion	The term Ozone depletion is used to describe two distinct, but related, observations: a slow, steady decline, of about 3% per decade, in the total amount of ozone in the earth's stratosphere during the past twenty years, and a much larger, but seasonal, decrease in stratospheric ozone over the earth's polar regions during the same period.
Fertilization	Fertilization is fusion of gametes to form a new organism. In animals, the process involves a sperm fusing with an ovum, which eventually leads to the development of an embryo.

CPSIA information can be obtained
at www.ICGtesting.com
Printed in the USA
BVHW012229170821
614670BV00014B/294

9 781428 803664